Proceedings in Life Sciences

Immunity in Invertebrates

Cells, Molecules, and Defense Reactions

Edited by Michel Brehélin

Editorial Assistants
J. M. Arcier, N. Boemare, J. R. Bonami
and C. P. Vivarès

With 72 Figures

Springer-Verlag
Berlin Heidelberg New York Tokyo

Dr. MICHEL BREHÉLIN
Laboratoire de Pathologie Comparée
USTL
Place E. Bataillon
F-34060 Montpellier

ISBN 3-540-15871-5 Springer-Verlag Berlin Heidelberg New York Tokyo
ISBN 0-387-15871-5 Springer-Verlag New York Heidelberg Berlin Tokyo

Library of Congress Cataloging in Publication Data. Main entry under title: Immunity in invertebrates. (Proceedings in life sciences) Reports from a conference held in Montpellier, France, Sept. 17–24, 1984. Bibliography: p. Includes index. 1. Invertebrates – Immunology – Congresses. I. Brehélin, M. (Michel), 1944–. II. Arcier, J. M. III. Series. QL364.I46 1985 592'.029 85-22056

Offsetprinting and bookbinding: Brühlsche Universitätsdruckerei, Giessen
2131/3130-543210

Preface

In 1822, E. Metchnikoff discovered phagocytosis while he was studying starfish larvae. The following year, while carrying out experiments on Daphnia (Crustacea), the same author demonstrated the scale of the phenomen in the defence of the organism. Finally, between 1891 and 1910, L. Cuénot was the first to reveal lymphoid organs, in crustaceans and insects; some of these organs play a role in both phagocytosis and in haematopoieses.

With rare exceptions, research into defence reactions in invertebrates was meagre in quantity and sometimes meagre in quality from then until the mid-1960's. The renewal of interest in the subject over the past few years is now well established, and invertebrates are no longer the poor relation of immunity research.

Three of the many reasons which have been found to account for this renewal of interest can be stressed. One concerns the preoccupations of fundamental research and the other two are associated with applied research. Firstly, there is hope of finding defence mechanisms in invertebrates that are new (or not yet known) or of using phylogenesis to find explanations for phenomena in vertebrates which are still unaccounted for. Secondly, the use of biological methods to control invertebrates which spread disease or which damage breeding operations can only be optimized if it is known how these invertebrates fight certain of their parasites or pathogenic agents. Finally, for obvious economic reasons, the development of intensive fish farming requires better knowledge of the phenomena which govern whether crustaceans and molluscs resist or are killed by the effect of potentially pathogenic germs.

This increase in the amount of research carried out necessitates better coordination between the various teams and the periodic holding of meetings at which scientists can discuss and compare their results. In addition, the great number of different lines of research and techniques used requires the publication of reports enabling every scientist to keep abreast of developments in fields in which he is not a specialist.

The International Society of Developmental and Comparative Immunology encourages the periodic holding of conferences in order to enable such meetings of scientists, and it was at the instigation of Professor E.L. Cooper, President of this society, that we organized the conference held in Montpellier from 17th–24th September 1984. We have used this meeting between sepcialists as an opportunity to publish the reports which compose the present volume.

This book does not contain exhaustive, chronological reviews of all the results published in each specialization. Our intention is to report on the state of progress of research and to set out clearly the knowledge acquired from various aspects of immunity in invertebrates. Each chapter is designed to provide information to keep students and nonspecialists up to date and to serve as a starting point for fresh research. Some of the reviews are from material given at the Montpellier conference. In order to obtain as comprehensive an approach as possible, we asked other scientists to write chapters on subjects that were not covered in depth at the conference.

The volume is divided into three sections. The first covers the immuno-competent cells. The second part contains the most recent findings in research on the molecules involved in defence reactions. The third section contains reviews of work on immune reactions involving the factors already defined. Work on the recognition of non-self, a large amount of which is being carried out at the moment, is the subject of several chapters and various opinions and theories are reported.

Publication has only been possible because all the colleagues asked agreed to devote some of their time to writing chapters. We hope that their work will now serve as a basis for scientists carrying out future work on immunity in invertebrates.

Montpellier, Autumn 1985

For the "Equipe de Recherche en Pathologie des Animaux Marins":

Dr. M. Brehélin

Acknowledgements. We would like to express our gratitude to the Life Sciences Department of the Centre National de la Recherche Scientifique (France) whose support made possible the conference in Montpellier and hence this book.

We would also like to thank Dr. D. Czeschlik from Springer-Verlag.

Contents

Contributors

You will find the addresses at the beginning of the respective contribution

Section 1 The Immunocompetent Cells in Invertebrates

Chapter 1 Insect Haemocyte Separation – An Essential Prerequisite to Progress in Understanding Insect Cellular Immunity

N. A. Ratcliffe[1], G. P. Mead[1], and L. R. Renwrantz[2]

CONTENTS

1 Introduction

There is an urgent need for purified populations of insect haemocytes in order to progress in our understanding of many areas of insect immunity. For example, recent work indicates that non-self recognition, and the subsequent cellular defences elicited, involve the interaction of two distinct blood cell populations [8,9]. The provision of definitive proof of this interaction would be greatly facilitated if purified populations of the reactive haemocytes were available for detailed studies in vitro. An outstanding recent report by Schlumpberger et al. [10] on the separation and labelling of subpopulations of the blood cells of the tunicate, *Botryllus* sp., also emphasizes the importance of pure cell preparations in studies on invertebrate immunity. These authors raised monoclonal antibodies against the *Botryllus* sp. blood cells and showed that these antibodies identified specific classes of morphologically distinct cell types. They also described the separation of these cell types by density gradient centrifugation and cell sorting. Having thus developed methods of clearly discriminating and separating the different blood cell types, Schlumpberger et al. [10] are now turning their attention to the functions of these blood cells at various stages of allodiscrimination between different colonies. In insects, a similar approach would be invaluable in resolving problems of haemocyte classification and cell lineage, and in clarifying apparent functional inconsistencies of the blood cells from one species to another.

---

1 Department of Zoology, University College of Swansea, Singleton Park, Swansea SA2 8PP, U.K.
2 Zoologisches Institut und Zoologisches Museum, Universität Hamburg, Martin-Luther-King-Platz 3, 2000 Hamburg 13, FRG

Immunity in Invertebrates
Ed. by M. Brehélin

2 Previous Work

Previous publications describing techniques for obtaining relatively pure populations of insect blood cells are confined to those by Peake [7], Bohn [2] and Chain and Anderson [3]. Of these reports, only that by Peake [7] was solely concerned with cell separation methodology and characterization of the separated cells. Peake [7] set up discontinuous gradients of Ficoll for the separation of haemocytes of *Calliphora vicina* and recovered a high percentage of prohaemocytes and phagocytes while the majority of the other cell types, including the oenocytoids and thrombocytoids, were lost by cell lysis or adhesion to the sides of the tubes. Detailed examination of Peake's [7] methodology indicates that its application would, however, be mainly confined to the Diptera. Many dipterans, unlike most other insects, do not appear to have the coagulation cell type, often referred to as the coagulocyte or cystocytes, and which induces a rapid haemolymph gelation. Instead, they contain peculiar anucleate, cell fragments, termed thrombocytoids, which rapidly agglutinate following wounding. Thus, only in the dipterans would it be possible to follow Peake's [7] methodology in which he bleeds the insects into a Tris saline, repeatedly resuspends the cells by suction through a Pasteur pipette and then centrifuges and resuspends the haemocytes, all presumably in endotoxin-containing and anticoagulant-free solutions, without inducing a massive and irreversible haemolymph gelation process (see Sect. 3).

Bohn [2], in his report on the differential adhesion of *Leucophaea maderae* haemocytes to glass surfaces, was mainly concerned with developing an in vitro system for studying insect immune phenomena rather than developing a technique for cell separation. He did, however, show that the two main haemocyte categories of *Leucophaea*, the plasmatocytes and granular cells, exhibited considerable differences in their adhesive properties and this allowed the production of nearly pure monolayers of either cell type. The problems with this separation technique are, however, that the defence responses may be triggered during adhesion to the glass and that the labile granular cells/cystocytes rarely remain intact but rapidly discharge their contents in vitro. Ideally, after cell separation, the blood cells should be in suspension, as in vivo, and thus readily available for subsequent manipulations.

Chain and Anderson's [3] paper on the selective depletion of *Galleria mellonella* plasmatocytes after bacterial injection, also describes methods for obtaining homogeneous cell types. Since, as with *Leucophaea*, the plasmatocytes showed stronger and more rapid adhesion onto glass, than the other cell types, highly enriched plasmatocyte preparations were obtained by allowing normal blood cells to spread on coverslips for 10–15 min before washing vigorously to remove all the other cells. Such preparations are suboptimal for subsequent manipulations, as indicated above for the *Leucophaea* haemocyte monolayers. Pure granular cell populations were, however, obtained by a different method, involving the injection of bacteria into the *Galleria* larvae before harvesting the cells. Chain and Anderson [3] showed that such injections resulted in plasmatocyte depletion, possibly by cell adhesion to the lining of the haemocoele, and the presence of nearly pure granular cells free in the haemolymph. This technique may well be useful, although the cells harvested are probably atypical due to triggering of the immune defences by the injected bacteria.

3 Problems in Cell Separation

There are a number of problems associated with separating the blood cells of insects, including problems during bleeding the animals, with the cell separation technique and with harvesting the cells.

3.1 Bleeding the Insects

Many insects, including most of the important vector species, are extremely small and therefore contain insufficient haemolymph for use in cell separation. However, and as detailed below, it may now be possible to pool the blood from a number of insects and at least partially overcome this difficulty.

Another formidable obstacle, as anyone who has worked with insects will realize, is the almost instantaneous coagulation reaction of the blood on exposure to air (see Chap. 14). In an attempt to discover an anticoagulant suitable for use with insects, we tested various techniques and a range of chemicals (Table 1), many of which are listed in Durliat and Vranckx's [4] paper on haemolymph coagulation in lobsters.

Initially, pre-cooled locusts were bled directly into syringes containing equiosmotic solutions of the test substances, and then the cystocytes were viewed for degranulation and lysis at various times after bleeding. The only satisfactory anticoagulant under these conditions was Carlson's saline with citrate and EDTA which stabilized the cells for up to 30 min (Table 1). In another series of tests, insects were exposed to glacial acetic acid vapour for 1–15 min before examining the cells. This technique was tried as in locusts, according to Grégoire [5], this stabilizes the coagulation cells for up to 3 h. The results did indeed show a stabilization of the cystocytes for 20–30 min in the three species tested but, frequently, numbers of cells recovered from the treated insects were significantly reduced. Thus cockroaches, which have much higher numbers of haemocytes than locusts, were utilized subsequently in many of these preliminary experiments. Finally, various combinations of acetic acid vapour exposure and injected anticoagulants were tested until a low pH saline containing citric acid and EDTA was evolved which stabilized the cystocytes for many hours in vitro [6]. These results show that the insect blood cells are considerably more unstable than the lobster [4] or other decapod [11] haemocytes which can be stabilized in vitro without the necessity of injecting anticoagulants. The use of endotoxin-free glassware and water is also an essential prerequisite for stabilizing the insect cells although these precautions are apparently unnecessary with crustacean blood cells (Smith and Söderhäll, pers. comm.). This is surprising considering that the prophenoloxidase system is not triggered by endotoxin in insects [1] (see Chap. 15).

Our insect anticoagulant, like that developed for crustaceans by Söderhäll and Smith [11], contains citric acid and EDTA which impart a low pH and abrogate the need to pretreat the insects in acetic acid vapour. For *Manduca sexta*, this anticoagulant consists of 0.098 M NaOH, 0.145 M Nacl, 0.017 M EDTA (free acid) and 0.041 M citric acid, and has a pH of 4.5 and an osmolality of 370 mOs kg^{-1}. This anticoagulant has also been successfully used, simply by adjusting the osmolality, with *Galleria mellonella* and *Blaberus craniifer* [6]. In the case of *Galleria,* sufficient numbers of

Table 1. Some of the techniques and chemicals tested for inhibiting the coagulation of insect haemolymph

	Species	Results
A. Without injection[a]		
NaCl + NEM[b]	*Locusta migratoria* *Schistocerca gregaria*	Cystocyte lysis within 5 min
NaCL + NEM + citrate[c]	*Locusta migratoria* *Schistocerca gregaria*	Massive cystocytic lysis by 15 min
Citrate EGTA or EDTA[d]	*Locusta migratoria* *Schistocerca gregaria*	Cystocyte lysis by 15 min in EGTA, greater stability of cells in EDTA up to 30 min
Carlson's + citrate + EDTA[e]	*Locusta migratoria* *Schistocerca gregaria*	Cystocytes stable up to 30 min but lysis thereafter
B. Glacial acetic acid vapour treatment[f]	*Locusta migratoria* *Schistocerca gregaria* *Blaberus craniifer*	Cystocytes stable up to 20–30 min but cell numbers may be reduced
C. With injection[g]		
NaCl + 5% cysteine	*Blaberus craniifer*	Cystocyte lysis/degranulation within 2–3 min
NaCl + citric acid + EDTA	*Blaberus craniifer* *Manduca sexta* *Galleria mellonella*	Cystocytes/granular cells stable for hours

[a] Chilled insects bled directly into ice-cold saline containing test substance
[b] NEM = 0.02 M N-ethyl-maleimide
[c] 10% Sodium citrate
[d] Formula in Söderhäll and Smith [11]
[e] Carlson's saline + 0.15 M sodium citrate + 0.38% EGTA or EDTA
[f] See Grégoire [5]
[g] 20 μl–1 ml per insect as described in Mead et al. [6]

blood cells for separation can be obtained by pooling the haemolymph from 5–10 larvae. Thus, the small size of an insect does not necessarily limit its use for haemocyte separation studies.

3.2 Cell Separation Techniques

Having stabilized the cells, there are many different techniques available for cell separation including density gradient centrifugation with Ficoll, Percoll or sodium metrizoate. Other methods involve velocity sedimentation, electrophoretic and chromatographic techniques. Most of these methods, however, are unsuitable with insect haemocytes, either due to the small volume of the blood available or as a result of the instability of the cells which would probably rapidly degranulate in contact with the columns or gels. The obvious choice for cell separation is Percoll (Pharmacia), which is a medium based on colloidal silica particles rendered non-toxic by coating with polyvinylpyrrolidone. Percoll also has a low osmolality, is compatible with living cells and can rapidly be used to form discontinuous gradients or continuous gradients following centrifugation.

For cell separation with continuous gradients of Percoll, it is, however, necessary to test a range of gradient concentrations from ca. 50–80% Percoll. By doing this, together with density marker beads, it is possible to generate gradients with different density profiles and to take maximal advantage of any density differences between the various cell types (see Pharmacia Instruction Manual).

Before forming the gradients and loading on the stabilized cells, however, the Percoll must be made isoosmotic to the haemolymph. Initially, we used NaCl to adjust the osmolality of the Percoll but found, in contrast to the crustacean haemostatic cells [11], that subsequently the cystocytes degranulated and lyzed during centrifugation. Attempts were therefore made to incorporate the anticoagulant, used for bleeding the insects, into the Percoll. This was unsuccessful as the mixture rapidly became opaque and milky due to precipitation of the Percoll at low pH. Another anticoagulant was therefore formulated with reduced citric acid concentration imparting a higher pH than the anticoagulant used for bleeding [6]. This solution was used, together with pyrogen-free water, to dilute the Percoll, and form gradients slightly hyperosmotic to the *M. sexta* blood. Percoll made up in this anticoagulant remains clear at 4°C for some hours. It does, however, rapidly precipitate at raised temperatures so that the stock Percoll should be made up just before use each day. The diluted Percoll was made slightly hyperosmotic to the insect haemolymph as trials showed that this also helped to stabilize the cells.

To form the continuous gradients, the diluted Percoll is centrifuged at 2°C for 20 min at 22,000 g. Finally, ca. 1 ml of haemolymph diluted in anticoagulant is carefully loaded on the preformed gradient and the whole re-centrifuged at 2°C for 20 min at 750 g to separate the cells.

This method with preformed continuous gradients of Percoll, as outlined above, is the technique now used routinely in our laboratory. Alternative ways of using Percoll are, however, available and include discontinuous and in situ gradients, both of which were tested. Before setting up discontinuous gradients, marker beads were incorporated into the continuous gradients to estimate the density of the various haemocyte types. The densities of the Percoll layers in the density gradients were then adjusted to those of the various haemocyte types so that they should have settled out in their corresponding density layer following centrifugation. Different densities of Percoll from 46–63% and from two- to eight-step gradients were tested with *M. sexta* but with little reproducible success. Occasionally, a greater than 95% purity of cells was achieved but usually the bands obtained at the various interfaces consisted of mixed populations of cells [6]. This may be due to the fact that the buoyant densities of the haemocytes often vary from one insect to another. In addition, the extent to which the blood cells have been stabilized, and degranulation prevented, may also greatly influence their buoyant densities and this will also vary from one preparation to another. Discontinuous gradients have, however, been used successfully in invertebrates such as *Botryllus* [10] in which rapidly degranulating, coagulation-type cells are wanting.

Finally, in situ gradients were also tested for separating insect blood cells. Instead of adding the haemocytes to the Percoll gradients after centrifugation, the cells were mixed with the appropriate Percoll dilution (60% and 70% tested) and then spun at 22,000 g for 20 min, in the hope that they would settle into the appropriate layer during gradient formation. Surprisingly, the haemocytes survived this treatment and formed a number of bands but these contained mixed cell types [6].

3.3 Harvesting the Cells

Using preformed continuous gradients of Percoll, as detailed above, the main haemocyte types of three different insect species i.e. *M. sexta, G. mellonella* and *B. craniifer,* have been separated (Fig. 1). With *Manduca*, 54% Percoll (as defined on p. 14 of the Pharmacia Handbook and in Ref. 6) gave maximal results while with *Galleria* and *Blaberus* 60% Percoll produced the best separation of the cells, although 80% Percoll was required to satisfactorily concentrate the *Galleria* spherule cells.

The separated haemocytes appeared as opaque bands in the gradients and various techniques were compared for harvesting the cells. These techniques included siphoning with capillary tubing from the top of the gradient, inserting hypodermic needles through the side or bottom of the centrifuge tube, and pipetting up the cell bands through the gradient with siliconized Pasteur pipettes. By using layers of density marker beads lying close together, it was shown that the most accurate method of sampling was with the Pasteur pipettes [6]. This technique is also extremely rapid and easy to use and was thus adopted for all subsequent experiments.

Cell recovery from the gradients was 50–55% of those loaded on, while harvested cells had a viability of nearly 100%. More recently, it has been possible to improve the percentage recovery to up to 65% by silicon-coating the centrifuge tubes to prevent cell adhesion. Another approach worth trying, and adopted by Peake [7] after he too only recovered 50% of the *Calliphora* cells from the gradients, is to load the cells on to the gradients halfway down the tube. Peake's cell recoveries increased from less than 50% to around 80% after he used this technique [7].

A final and extremely serious problem relates to the necessity of washing and re-centrifuging the harvested cells to remove the anticoagulant and concentrate the separated cell populations. Even though the centrifuge tubes were silicon-coated and a cushion of Percoll was included in the bottom of the tubes, only 20–30% of the harvested cells were recovered after washing [6]. This means that only about one-tenth

Fig. 1 A-D. The main haemocytes types of *Galleria mellonella* before (plate **A**), and after (plates **B-D**) separation on continuous Percoll gradients. The haemocytes before separation are in an anticoagulant solution which limits the spreading of plasmatocytes (*Pl*). *G. mellonella* haemolymph is composed of ca. 47% plasmatocytes (*Pl*) and ca. 48% granular cells (*Gr*) but the other cell types, namely, the spherule cells (*Sph*) and oenocytoids (*Oen*), are also present in the plate. The haemocytes which have been separated (plates **B-D**) have also been washed and concentrated in a NaCl solution. *G. mellonella* granular cells (plate **D**) usually appear highly refractile and are readily distinguishable from plasmatocytes (plate **B**). The spherule cells (plate **C**) form only ca. 3% of the haemocytes and it is more difficult to obtain a concetrated preparation of this cell type. Oenocytoids only form ca. 1% of the blood cells and could not be reproducibly recovered from the gradients used. All scale *bars* are 10 μm

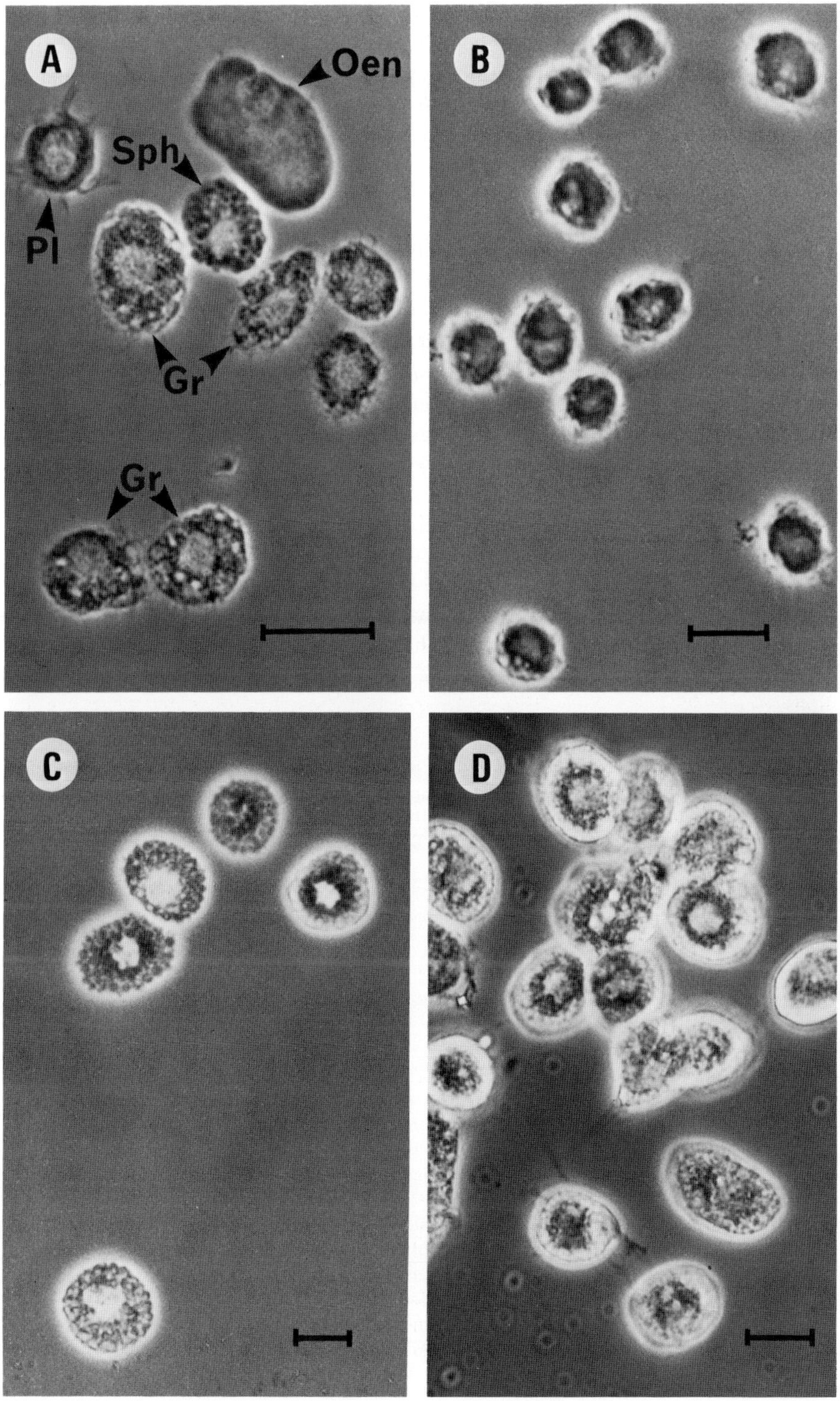
A
Oen
Sph
Pl
Gr
Gr
B
C
D

of the cells loaded on to the gradients are recovered after this second centrifugation. Insect blood cells are obviously extremely sensitive to centrifugation after the initial separation procedure. Attempts to improve cell recovery by loading more than ca. 3×10^6 cells on to each gradient always failed. In order to obtain sufficient cells for experimentation it should, however, be possible to increase the percentage harvested and to set up multiple gradients for pooling the purified cells.

4 Concluding Remarks

A method for separating the blood cells of insects has been successfully evolved. Many problems have been encountered during the development of this procedure due to the fragile nature of the cells. Successful separation of the insect haemocytes into bands on continuous Percoll gradients has, however, been achieved in 75–80% of the trials. The level of cell purity in *M.sexta* is in excess of 90% for the granular cells and 80% for the plasmatocytes. The main problem still remaining concerns the low level of cell recovery following washing of the haemocytes to remove the anticoagulant and concentrate the cells. Simple modifications such as silicon-coating the centrifuge tubes used for forming the gradients and testing insects at different larval stages will undoubtedly overcome this difficulty. Meanwhile, the purified haemocytes are being utilized to study the distribution of the prophenoloxidase system in the various cell populations and the interactive nature of the insect cellular and humoral defence reactions.

Acknowledgements. We are grateful to Dr. S.E. Reynolds of the University of Bath for supplying the *M. sexta* eggs and to Mrs. M. Colley for help in rearing the insects used in this research. The work was supported by the Natural Environment Research Council (G.P.M.), the Science and Engineering Research Council (grant numbers GR/B/60958 and GR/D/21684) and the Royal Society. The British Council and the Royal Society also kindly provided travel funds to L.R.R. and N.A.R.

References

1. Ashida M, Ishizaki Y, Iwahana H (1983) Activation of prophenoloxidase by bacterial cell walls or β-1,3-glucans in plasma of the silkworm, *Bombyx mori*. Biochem Biophys Res Commun 113: 562-568
2. Bohn H (1977) Differential adhesion of the haemocytes of *Leucophaea maderae* (Blattaria) to a glass surface. J Insect Physiol 23: 185-194
3. Chain BM, Anderson RS (1982) Selective depletion of the plasmatocytes in *Galleria mellonella* following injection of bacteria. J Insect Physiol 28: 377-384
4. Durliat M, Vranckx R (1981) Action of various anticoagulants on hemolymphs of lobsters and spiny lobsters. Biol Bull (Woods Hole) 160: 55-68
5. Grégoire C (1970) Haemolymph coagulation in arthropods. Symp Zool Soc Lond 27: 45-74
6. Mead GP, Ratcliffe NA, Renwrantz LR (1985) The separation of insect haemocyte types on Percoll gradients: methodology and problems. J Insect Physiol (in press)

7. Peake PW (1979) Isolation and characterization of the haemocytes of *Calliphora vicina* on density gradients of Ficoll. J Insect Physiol 25: 795-803
8. Ratcliffe NA, Rowley AF (1979) Role of hemocytes in defense against biological agents. In: Gupta AP (ed) Insect hemocytes. Development, forms, functions and techniques. Cambridge University Press, New-York, pp 331-414
9. Ratcliffe NA, Leonard C, Rowley AF (1984) Prophenoloxidase activation: nonself recognition and cell cooperation in insect immunity. Science 226: 557-559
10. Schlumpberger JM, Weissmann IL, Scofield VL (1984) Separation and labelling of specific subpopulations of *Botryllus* blood cells. J Exp Zool 229: 401-411
11. Söderhäll K, Smith VJ (1983) Separation of the hemocyte populations of *Carcinus maenas* and other marine decapods, and prophenoloxidase distribution. Dev Comp Immunol 7: 229-239

Chapter 2 Surface Membrane Components of Circulating Invertebrate Blood Cells and Their Role in Internal Defense

T. P. Yoshino[1]

CONTENTS

1 Introduction

For eukaryotic cells the importance of plasma membrane constituents in conveying chemical information regarding the extracellular environment is well established. Specific immune recognition and effector/regulatory functions in higher vertebrates are dependent upon a variety of immunocyte-associated membrane factors which include receptors for major histocompatibility (MHC) antigens [44], antigen-specific T-cell receptors [28,40], membrane immunoglobulins [71], Fc and complement receptors [13,17,54], hormone receptors [43], and glycosyltransferases [19]. Extensive knowledge of these cell surface components has been instrumental in helping to provide an understanding of the complex cellular and molecular interactions involved in the generation of immune responses in these animals.

In contrast to what is known regarding the biochemical basis for cellular reactivity in the vertebrate immune system, little information is available for invertebrates, in particular the role of blood cell surface membrane components in regulating cellular interaction with foreign materials. Although lacking the classical immune recognition molecules of vertebrates (surface Ig, T-cell MHC or antigen receptors), motile blood elements and fixed phagocytic cells of most metazoan invertebrates are still highly capable of distinguishing between "self" and "non-self" [15]. It has long been assumed that specific surface membrane constituents were responsible for the discriminatory

1 Department of Zoology, University of Oklahoma, 730 Van Vleet Oval, Norman, Oklahoma 73019, USA

Immunity in Invertebrates
Ed. by M. Brehélin

capacity of these invertebrate cells, although too little information regarding the chemical make-up of integral or cytophilic membrane structures has hindered the critical evaluation of their role in mediating invertebrate immunocyte reactions. The purpose of this report is to review evidence supporting a functional role of selected blood cell membrane components in facilitating internal defense responses, and to describe the various approaches being used in the molecular mapping of these and other surface membrane structures.

2 Evidence for Receptor-Mediated Blood Cell Functions

Differential reactivities of circulating blood elements to introduced foreign materials have been documented in numerous studies of phagocytosis and cytotoxicity employing a variety of invertebrate species (see reviews [4,15,34,46]). Undoubtedly, differences in surface charge or hydrophobic properties of foreign materials could account for some of the variation in cellular reactivity [35,42]. However, there is considerable evidence that certain blood cell-foreign body reaction's may be receptor-mediated (i.e., are dependent upon the specific biochemical interaction of blood cell plasma membrane components with foreign molecular groupings).

2.1 Phagocytosis Studies

In vitro phagocytosis of foreign particles by circulating blood elements has been shown to be at least partially dependent on opsonic factors occurring in homologous serum (cell-free hemolymph) [22,27,32,49,60,65], suggesting the presence of receptors for these endogenous opsonizing agents on phagocyte surface membranes. Little is known regarding the macromolecular structure of membrane opsonin receptors, except that in cases where carbohydrate-binding agglutinins (lectins) may be serving as opsonic factors [27,49,60], membrane glycoproteins, glycolipids, or polysaccharides are assumed to be the source of sugar moieties serving as specific lectin-binding sites. Moreover, the observation that phagocytosis of target particles in some species can proceed in the complete absence of humoral factors [7,20,53,62] indicates the possible existence of constituitive or cytophilic plasma membrane components capable of directly interacting with foreign materials. Renwrantz and Stahmer [49] have recently shown that Ca^{2+} alone could stimulate in vitro phagocytosis of yeast by *Mytilus* hemocytes in the absence of serum opsonins, and have hypothesized the presence of a calcium-dependent, membrane-associated recognition factor to account for their observation. Although the exact chemical nature of this membrane factor(s) is still unknown, they demonstrated further that antibodies raised against a purified serum opsonin (an agglutinin) cross-reacted with membrane components of washed hemocytes indicating that these cells are exhibiting at their surface, molecules with structural similarities to the circulating opsonin [49]. These results, combined with an earlier finding that a serum opsonin/agglutinin in the snail, *Lymnaea stagnalis* [31], also occurs as a hemocyte membrane-bound constituent argues strongly for endogenous membrane agglutinins serving as primitive recognition factors in these animals (see Chap. 9). Moreover, a similar co-occurrence of soluble and hemocyte-associated lectins in lepidopteran insects [2], blue crabs [69], and oysters [70] further suggests that such a recognition pathway may represent a more generalized one in various groups of invertebrates.

Without deemphasizing the importance of agglutinins as potential immunorecognition molecules in the invertebrates, it should be pointed out that other nonagglutinin factors may also be serving a similar function in some species. Circulating phagocytes from echinoderms appear to possess surface receptors for mammalian complement fragment C3bi, and possibly C3b, as demonstrated by an uptake of C3b and C3bi-coated erythrocytes (EAC1243b, EAC43b, EAC4bi) at levels significantly above erythrocytes coated with other complement components, antibody alone, or with no coating [9]. The subsequent finding that complement-like substances in the coelomic fluid of this same animal [10] also promoted erythrophagocytosis by echinoid and mammalian phagocytes prompted the intriguing speculation that the complement lytic and opsonizing system of vertebrates evolved early in the dueterostome line and has since been conserved with some molecular modification during vertebrate phylogenesis. Although protosomes appear to lack the complement-like opsonizing system of echinoderms, they also possess opsonins which apparently lack agglutinating activity [22, 66]. In arthropods, activation of the prophenoloxidase pathway by microbial products appears to mediate a non-self recognition process as manifested by enhanced phagocytosis or degranulation by hemocytes in vitro [45,63]. Moreover, in the wax moth, *Galleria mellonella*, two distinct blood cell-types appear to work cooperatively in producing an effective phagocytic response [45]. Prophenoloxidase-positive granular hemocytes initiate the recognition process by releasing the enzyme which is activated at the microbial cell surface resulting in the coating of the target organism with products of the phenoloxidase cascade. A second type of leukocyte, the prophenoloxidase-negative plasmatocyte, is then able to recognize coated microbes as foreign and respond by enhanced phagocytic activity. Hemocyte receptors of foreignness may be involved in either the initial degranulation reaction or in the microbial endocytotic phase, although to date such receptors have not yet been identified.

2.2 Cytotoxicity Studies

Cytotoxic reactions, likewise, appear to be receptor-mediated. Decker et al. [18] have shown that circulating blood elements from representatives of several invertebrate phyla (molluscs, arthropods, echinoderms) exhibit a natural cytotoxicity toward a variety of vertebrate cell targets using a standard Cr-release cytotoxicity assay. Target cell killing could be inhibited in the presence of specific mono- or disaccharides, suggesting that foreign, target recognition and the activation of cytotoxic processes are mediated through carbohydrate-reactive components on blood cell surfaces. Again, membrane lectins appear to be serving as target cell receptor sites.

In another example using a naturally occurring parasite-host system, Bayne and coworkers [5] have demonstrated that serum-free hemocytes from a *Schistosoma mansoni* (blood fluke)-resistant strain of the snail, *Biomphalaria glabrata*, are cytotoxic for larval parasites in vitro. Membrane-bound cytotoxic factors on resistant snail hemocytes are also thought to be cytophilic in nature since exposure of hemocytes from a susceptible *B. glabrata* strain to resistant snail serum in vitro transforms these normally harmless cells into potent "killer" hemocytes [6,37]. A partial protection produced in susceptible snails passively inoculated with resistant snail serum 24 h prior to or 24 h after exposure of susceptible hosts to infective larval stages further suggests that the cytotoxic transformation observed in vitro may also be occurring in the intact animal [25].

3 Molecular Probe Analyses of the Blood Cell Surface

Although many of the previously cited studies provide evidence that specific components associated with blood cell surface membranes are responsible for the ability of these cells to discriminate "self" from "non-self" and generate appropriate responses, the identity, origin, and structure of these membrane constituents are still largely unknown. Even in the case of the well characterized lectin-like opsonins, the chemical nature of the membrane macromolecules serving as the cell-bound opsonin or as receptors for humoral opsonic factors has not been fully elucidated.

One approach to investigating the invertebrate blood cell surface has been to utilize a variety of probe reagents to identify and characterize the molecules comprising the external plasma membrane of circulating cells. Using molecular probes, several broad groups or categories of membrane components have now been recognized including receptors for mammalian erythrocytes, lectin-binding determinants, and surface antigens defined by their reactivity with specific probe antibodies. The overall objective of this type of study has been to initiate the mapping of endogenous surface membrane constituents of circulating blood elements. Defined surface markers might then be useful in analyzing functionally distinct cellular subpopulations or in identifying surface components which share structural similarities with immunocytes of vertebrate origin. The following is a brief review of representative investigations.

3.1 Erythrocyte Receptors

The spontaneous attachment of vertebrate erythrocytes (RBC's) to the surface of invertebrate blood cells forming flower-like rosettes provides a useful technique for identifying the presence of specific surface receptors for target cells. Due to an earlier finding that receptors for sheep erythrocytes (SRBC's) could be used as a defined surface marker for mammalian lymphocytes [29], great interest was generated when a subpopulation of nonadherent coelomocytes from earthworms were also shown to possess surface receptors for SRBC's [16,67]. The morphological and behavioral resemblance of rosetting cells to vertebrate lymphocytes, the stable nature of spontaneous rosettes, and the coelomocyte binding specificity for sheep, but not human or mouse RBC's prompted speculation of a possible phylogenetic link between these two celltypes [67]. However, as pointed out by these investigators, one must be cautious in drawing too rapid a conclusion without a direct biochemical comparison of the isolated receptor. It is now apparent that surface receptors for RBC's from sheep and other vertebrates are widespread in a variety of invertebrate species, including insects [3], crustaceans [39], and molluscs [48], as well as on vertebrate cells of non-lymphoid origin [36], and, therefore, the phyletic relationship (if any) between erythrocyte receptors on invertebrate and vertebrate immunocytes still remains unclear.

3.2 Lectin-Binding Determinants

Heterologous lectins (primarily of plant origin) with defined sugar-binding specificities have been extensively used as membrane probes to analyze the glycoconjugates composing the surface membranes of blood cells from a variety of invertebrate species, including molluscs [14,47,56,57,61,73,79], annelids [51], and arthropods [21]. In some cases, subpopulations of circulating cells could be defined by the occurrence of distinctive sugar reactive sites. For example, nonadherent leukocytes of the earthworm, *Eisenia foetida*, possess receptors for concanavalin A (Con A), whereas similar receptors on adherent cells are largely absent [51]. Similarly, adherent granular hemocytes of planorbid snails appear to react uniquely with Con A and galactosyl-specific *Ricinus* agglutinin, while adherent agranular cells are nonreactive to these same lectins [56]. It is clear from studies such as these that blood cell populations can exhibit distinctive interspecific lectin binding specificities, as well as differences in the expression of binding specificity among cells of a single individual. In addition, lectin probes may be useful in characterizing the functional state of blood cells involved in immune reactions. In *Drosophila* (Tum^1 mutant), the capacity of the host to effectively encapsulate and kill eggs of a parasitic wasp was associated with a high prevalence of wheat germ agglutinin-reactive hemocytes (see Chap. 13), whereas under conditions in which fly larvae were rendered susceptible to parasitism, the prevalence of WGA^+ hemocytes was low [41]. In this example, host immune competence was directly correlated to a cell subpopulation expressing WGA-reactive surface determinants.

Functional analogies in the response of vertebrate and invertebrate immunocytes to lectin binding suggest that the chemical "messages" imparted to cells of widely disparate phylogenies by these ligands may be quite similar. These responses include the stimulation of mitogenesis by Con A, PHA, pokeweed mitogen, or LPS in earthworms [52,68] and starfish [12], the redistribution and internalization of lectin-receptor complexes in earthworm coelomocytes [51], and the hemocytes of various molluscan species [73,74,79], lectin-mediated endocytosis by snail hemocytes [56,58], and, also in snails, lectin-induced in vitro cytotoxic reactions against larval trematodes by normally benign hemocytes [11] in a fashion analogous to lectin-dependent cell-mediated cytotoxic reactions of mammalian lymphocytes [8]. Although it should be kept in mind that the above investigations employed lectins of heterologous origin, these studies make it clear that the specific chemical interaction of invertebrate surface components with external ligands can trigger responses which, in many cases, could be directly related to internal defense processes. Further investigation of lectin-mediated events should continue to provide important information regarding the basic mechanisms of chemical reception and responsiveness in these cells.

3.3 Blood Cell Surface Antigens

The application of polyclonal and monoclonal antibodies as molecular probes of both structure and function is well established and represents one of the basic tools in cell biological research. The use of antibodies in the study of invertebrate blood cells has grown in recent years and has contributed significantly to our understanding of their surface chemistry.

Two primary approaches have been used to serologically investigate circulating invertebrate cells; one approach has been to produce antibodies directly against the blood cells to be analyzed or to a related substance (e.g., a serum factor) for use in membrane probe studies (homologous antibody approach). The other is to use antibodies raised against defined antigens from an exogenous source, and to determine whether these antibodies will cross-react with blood cell components thereby establishing a serological relatedness (i.e., structural similarity) between the defined antigen and leukocyte surface structures (heterologous antibody approach). Although caution must always be exercised when inferring protein structural relatedness through serological cross-reactivity [38], the application of antibody probes to invertebrate cell research has contributed greatly to our understanding of the molecular make-up of blood cell membrane constituents.

3.3.1 Homologous Antibody Studies

The use of polyclonal antibodies to soluble hemolymph components as membrane probes has provided considerable evidence that serum and circulating cells share many of their molecules. Several examples have already been cited in which blood cell surface components have been shown to be structurally related to serum agglutinins or lectins [31,49,70] and appear to be serving as cell-bound recognition factors mediating attachment of foreign material to the blood cell surface [31,32,49]. However, the functional significance of other plasma membrane molecules is not as clear. For example, antibodies generated against soluble hemoglobin of the snail, *Biomphalaria glabrata*, strongly react with surface components associated with the snail's hemocytes [75]. It has been suggested that another respiratory pigment in molluscs, hemocyanin, may possess opsonic properties [30,66], and, if so, perhaps one of the functions of hemocyte-associated hemoglobin is as a recognition factor.

The use of monoclonal antibodies (mAB's) to dissect the antigenic heterogeneity exhibited by invertebrate cell surfaces has added an important dimension to our understanding of the molecular make-up of circulating blood elements from a cell population perspective. Recently it has been reported that blood cells of the colonial tunicate, *Botryllus* sp., are composed of at least three antigenic subpopulations of cells based on their reactivity with a panel of six mAB's [55] and, in an earlier study, Yoshino and Granath [77] similarly demonstrated that the adherent population of circulating snail hemocytes was actually composed of several antigenically distinct cell subpopulations based on the presence of mAB reactive epitopes and the patterns of antigen distribution on hemocyte surface membranes. In a related study [78], they further demonstrated that subpopulation-specific mAB's could be very useful in deter-

Table 1. Phagocytic (Phag) and acid phosphatase (APase) activities of *Biomphalaria glabrata* hemocytes in relation to the expression of the BGH_1 cell surface antigen [78]

Hemocyte subpopulation	Phagocytic activity		Acid phosphatase activity	
	Phag$^+$	Phag$^-$	APase$^+$	APase$^-$
BGH_1^+	1.9 ± 0.9[a]	36.4 ± 9.4	4.6 ± 1.9	29.1 ± 3.8
BGH_1^-	17.3 ± 4.8	44.4 ± 11.7	63.1 ± 4.3	3.2 ± 1.1

[a] % ± SD; N = 10 snails per assay

mining whether antigenic subsets might also be exhibiting functional differences. One subpopulation of *B. glabrata* hemocytes expressing a surface epitope, designated BGH_1, was found to be significantly less active in the phagocytosis of test particles and in the production of the lysosomal hydrolase, acid phosphatase, than BGH_1^- cells (Table 1). In addition it was shown that the prevalence of the weakly-phagocytic BGH_1^+ cells was consistently higher in a strain of *B.glabrata* (PR albino, M-line) which was highly susceptible to larval schistosome parasites (PR-1 strain) when compared to hemocytes from a schistosome-resistant snail strain (10-R2) (Fig. 1). This observation, combined with an earlier finding that resistant 10-R2 snails possessed twice the concentration of circulating cells as the M-line strain [23], prompted speculation that natural susceptibility or resistance in these hosts might be directly related to the number of circulating highly reactive BGH_1^- hemocytes [78].

Results of a previous study showed that natural infection by larval schistosomes induced an acute leukocytosis in M-line snails [24] indicating that large numbers of cir-

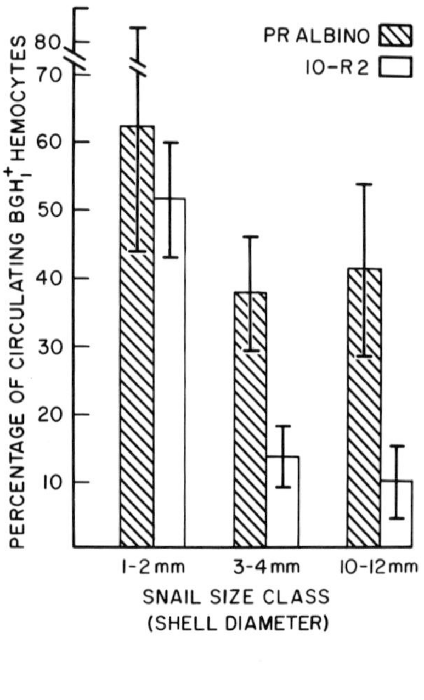

Fig. 1. A comparison of the prevalence of circulating BGH_1^+ hemocytes found in three size classes of two *Biomphalaria glabrata* strains (*PR albino* and *10-R2*). The N for each size class was 10, 8, and 15 for 1–2 mm, 3–4 mm, and 10–12 mm snails, respectively. *Vertical bars* represent SD [78]

culating cells are mobilized in susceptible snails upon parasite entry. If BGH_1^- hemocytes were indeed the primary effector cell, it might be predicted that the composition of newly recruited circulating hemocytes in susceptible snails would be of the weakly-phagocytic BGH_1^+ type. To test this prediction, 40 snails of the M-line stain and 40 of the 10-R2 strain were exposed individually to 10 *S. mansoni* miracidia each and duplicate hemolymph samples were taken from 10 randomly chosen snails of each strain at 0 h (unexposed), 1 h postexposure (PE), 12 h PE, and 24 h PE. The cells in one blood sample were counted to determine the concentration of circulating hemocytes and the other sample was subjected to fluorescent antibody analysis [77] to assess the proportion of BGH_1^+/BGH_1^- hemocytes in each snail at each time interval. As shown in Fig. 2, parasite infection of susceptible M-line snails induces a fourfold increase in the number of circulating hemocytes 1 h PE, followed by a decrease to control levels (0 h) by 12 h PE ($F[3,36] = 18.80, p < 0.001$). Circulating hemocyte levels in infected 10-R2 snails decrease at 1 h PE ($F[3,36] = 4.97, p < 0.01$), presumably due to the removal of cells from the circulation to participate in encapsulation reactions. However, contrary to our prediction, analysis of the antigenic subpopulations composing the circulating cells of M-line snails reveals that at 1 h PE the number of highly reactive BGH_1^- hemocytes significantly increases to levels above those found in resistant 10-R2 snails ($F[3.36] = 22.08, p < 0.001$) (Fig. 3). This latter result, combined with the finding that circulating 10-R2 BGH_1^- hemocytes, but not BGH_1^+ cells, decrease at 1 h PE (Fig. 3) indicates that in the 10-R2 snail stain hemocytes lacking the BGH_1 epitope most probably represent the primary effector cell involved in mediating parasite resistance. However, since infection of the susceptible

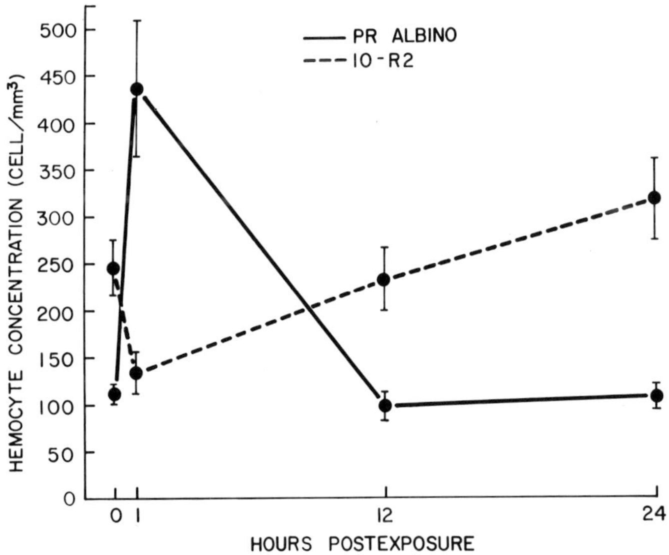

Fig. 2. Concentration of circulating hemocytes (all cells) from two strains of the snail, *Biomphalaria glabrata*, at various times postexposure to miracidia of *Schistosoma mansoni*. *Vertical bars* represent SD

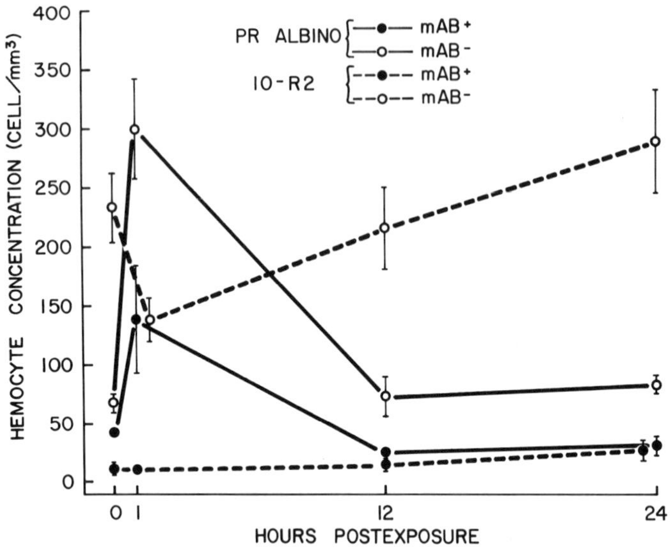

Fig. 3. Concentration of mAB⁺ and mAB⁻ hemocytes from two strains of *Biomphalaria glabrata* at various times postexposure to larval *Schistosoma mansoni*. Antibody-reactive and nonreactive cell numbers were determined by the following formula: total number of circulating hemocytes (Fig. 2) x prevalence (%) of mAB⁺ or mAB⁻ cells at each time PE and for each host strain. MAB⁺ hemocytes denote those expressing the BGH_1 surface epitope. *Vertical bars* represent SD

M-line strain elicits a strong BGH_I^- leukocytosis which surpasses cell concentrations found in 10-R2 snails, the number of highly reactive BGH_I^- cells per se apparently is not an absolute requirement in determining resistance to larval schistosome infection.

3.3.2 Heterologous Antibody Studies

Antibodies raised against a variety of defined heterologous antigens have been used to probe the surface membrane of invertebrate blood cells. The primary goal in such analyses has been to identify in invertebrates, molecules which display structural similarities to their vertebrate molecular counterparts, with the ultimate aim of understanding the phylogenetic origins of vertebrate immune factors. Interest has focused primarily on the molecules which make up the immunoglobulin superfamily, since it was first reported that a Thy-1-like molecule occurred in squid which shared amino acid sequence homology with mammalian Thy-1 and immunoglobulin V_λ domains [72]. Serological cross-reactivity of antibodies to highly purified mouse Thy-1 has subsequently been reported to occur on blood cells of snails [76]. In addition, serological evidence has also been produced which indicates that earthworm leukocytes express membrane components with structural similarities to human B_2-microglobulin [50], another member of the Ig superfamily found noncovalently associated with cell surface MHC antigens [72]. The earlier finding of B_2-microglobulin-like activity in extracts of a variety of invertebrates [59] suggests that this molecule or class of molecule may be widespread in the animal kingdom.

Fibronectin represents another substance which appears to be shared between vertebrates and invertebrates. Substances serologically related to human fibronectin have been reported in developing sea urchin embryos [64], colonial Porifera [33], and on snail hemocytes [76], although its function in these animals is still unknown. Possibly it functions in cell-cell adhesion or in the stabilizing of extracellular matrices as has been shown in vertebrates [1]. Moreover, with the recent finding that fibronectin can serve as an effective opsonin for mammalian macrophages [26], one cannot rule out a possible immunobiological function for similar substances in invertebrates.

Acknowledgments. The author would like to thank Dr. Carl Boswell for his help in reviewing this manuscript. The original data reported in this paper resulted from research supported by Grant AI 15503-07 from the NIH, NIAID, and Research Career Development Award AI 00491-01A1 from the same institute.

References

1. Akiyama SK, Yamada KM, Hayashi M (1981) The structure of fibronectin and its role in cellular adhesion. J Supramol Struct Cell Biochem 16: 345-358
2. Amirante GA (1976) Production of heteroagglutinins in haemocytes of *Leucophaea maderae* L. Experientia (Basel) 32: 526-528
3. Anderson RS (1977) Rosette formation by insect macrophages. Inhibition by cytochalasin B. Cell Immunol 29: 331-336
4. Bayne CJ (1983) Molluscan immunobiology. In: Saleuddin ASM, Wilbur KM (eds) Biology of Mollusca. Academic Press, New York, pp 407-486
5. Bayne CJ, Buckley PM, DeWan PC (1980) Macrophage-like hemocytes of resistant *Biomphalaria glabrata* are cytotoxic for sporocysts of *Schistosoma mansoni* in vitro. J Parasitol 66: 413-419
6. Bayne CJ, Buckley PM, DeWan PC (1980) *Schistosoma mansoni*: Cytotoxicity of hemocytes from susceptible snail hosts for sporocysts in plasma from resistant *Biomphalaria glabrata*. Exp Parasitol 50: 409-416
7. Bayne CJ, Moore MN, Carefoot TH, Thompson RJ (1979) Hemolymph functions in *Mytilus californianus*: The cytochemistry of hemocytes and their responses to foreign implants and hemolymph factors in phagocytosis. J Invertebr Pathol 34: 1-20
8. Berk G (1983) Cytotoxic T-lymphocytes. How do they function? Immunol Rev 72: 5-42
9. Bertheussen K (1982) Receptors for complement on echinoid phagocytes II. Purified human complement mediates echinoid phagocytosis. Dev Comp Immunol 6: 635-642
10. Bertheussen K (1983) Complement-like activity in sea urchin coelomic fluid. Dev Comp Immunol 7: 21-31
11. Boswell CA, Bayne CJ (1985) *Schistosoma mansoni*: Lectin-dependent cytotoxicity by hemocytes from susceptible host snails *Biomphalaria glabrata*. Exp Parasitol 60: 133-138
12. Brillouet C, Leclerc M, Panijel J, Binaghi R (1981) In vitro effect of various mitogens on starfish (*Asterias rubens*) axial organ cells. Cell Immunol 57: 136-144
13. Capron M, Capron A, Dessaint JP, Torpier G, Gunnar S, Johansson O, Prin L (1981) Fc receptors for IgE on human and rat eosinophils. J Immunol 126: 2087-2092
14. Cheng TC, Huang JW, Karadogan H, Renwrantz LR, Yoshino TP (1980) Separation of oyster hemocytes by density gradient centrifugation and identification of their surface receptors. J Invertebr Pathol 36: 35-40
15. Chorney MJ, Cheng TC (1980) Discrimination of self and non-self in invertebrates. In: Marchalonis JJ, Cohen N (eds) Self/non-self discrimination. Plenum, New York, pp 37-54 (Contemporary Topics in Immunobiology, vol 9)

16. Cooper EL (1973) Evolution of cellular immunity. In: Braun W, Ungar J (eds) Non-specific factors influencing host resistance. Karger, Basel, pp 11-23
17. Cunningham-Rundles C, Siegal FP, Good RA (1978) Isolation and characterization of a human mononuclear cell Fc receptor. Immunochemistry 15: 365-370
18. Decker JM, Elmholt A, Muchmore AV (1981) Spontaneous cytotoxicity mediated by invertebrate mononuclear cells toward normal and malignant vertebrate targets: Inhibition by defined mono- and disaccharides. Cell Immunol 59: 161-170
19. Endres RO, Lucas DO (1977) Lymphocyte surface glycosyltransferases. In: Regulatory mechanisms in lymphocyte activation. Academic Press, New York, pp 423-425
20. Gagen SJ, Ratcliffe NA (1976) Studies on the in vivo cellular reactions and fate of injected bacteria in *Galleria mellonella* and *Pieris brassicae* larvae. J Invertebr Pathol 28: 17-24
21. Gebbinck J (1980) Some observations on concanavalin A receptors on free cells of marine invertebrates. Dev Comp Immunol 4: 33-42
22. Goldenberg PZ, Greenberg AH (1983) Functional heterogeneity of carbohydrate-binding hemolymph proteins: Evidence of a nonagglutinating opsonin in *Homarus americanus*. J Invertebr Pathol 43: 33-41
23. Granath WO Jr, Yoshino TP (1983) Characterization of molluscan phagocyte subpopulations based on lysosomal enzyme markers. J Exp Zool 226: 205-210
24. Granath WO Jr, Yoshino TP (1983) Lysosomal enzyme activities in susceptible and refractory strains of *Biomphalaria glabrata* during the course of infection with *Schistosoma mansoni*. J Parasitol 69: 1018-1026
25. Granath WO Jr, Yoshino TP (1984) *Schistosoma mansoni*: Passive transfer of resistance by serum in the vector snail, *Biomphalaria glabrata*. Exp Parasitol 58: 188-193
26. Gudewicz PW, Molnar J, Lai MJ, Beezhold DW, Siefring GE, Credo RB, Lorand L (1980) Fibronectin-mediated uptake of gelatin-coated latex particles by peritoneal macrophages. J Cell Biol 87: 427-433
27. Hardy SW, Fletcher TC, Olafsen JA (1977) Aspects of cellular and humoral defense mechanisms in the Pacific oyster, *Crassostrea gigas*. In: Solomon JB, Horton JH (eds) Developmental immunobiology. Elsevier/North Holland, Amsterdam, pp 59-68
28. Janeway CA Jr (1980) Idiotypes, T-cell receptors, and T-B cooperation. In: Marchalonis JJ, Cohen N (eds) Self/non-self discrimination. Plenum, New York, pp 171-203 (Contemporary Topics in Immunobiology, vol 9)
29. Jondal M, Holm G, Wizzel J (1972) Surface markers on human T and B lymphocytes I. A large population of lymphocytes forming nonimmune rosettes with sheep red blood cells. J Exp Med 136: 207-215
30. Knaap WPW van der (1981) Recognition of foreignness in the internal defence system of the freshwater gastropod *Lymnaea stagnalis*. In: Solomon JB (ed) Aspects of developmental and comparative immunology. Plenum, Oxford, pp 91-97
31. Knaap WPW van der, Boerrigter-Barendsen LH, Hoeven DSP van den, Sminia T (1981) Immunocytochemical demonstration of a humoral defence factor in blood cells (amoebocytes) of the pond snail, *Lymnaea stagnalis*. Cell Tissue Res 219: 291-296
32. Knaap WPW van der, Sminia T, Schutte R, Boerrigter-Barendsen LH (1983) Cytophilic receptors for foreignness and some factors which influence phagocytosis by invertebrate leucocytes: in vitro phagocytosis by amoebocytes of the snail *Lymnaea stagnalis*. Immunology 48: 377-383
33. Labat-Robert J, Ravans de Ceccatty M, Robert L, Auger C. Lethias C, Garrone R (1979) Surface glycoproteins of sponge cells: Presence of a fibronectin-like protein on differentiated sponge cell membranes; its role in cell aggregation. In: Schauer T, Boer P, Buddecke E, Kramer MF, Vliegenthart JFG, Weigandt H (eds) Glycoconjugates. Thieme, Stuttgart, pp 431-432
34. Lackie AM (1980) Invertebrate immunity. Parasitology 80: 393-412
35. Lackie AM (1983) Effect of substratum wettability and charge on adhesion in vitro and encapsulation in vivo by insect haemocytes. J Cell Sci 63: 181-190
36. Lalezari P, Nehlsen SL, Sinha SBP, Stemerman MB, Veith FJ (1974) Spontaneous macrophage-erythrocyte rosette formation.A species-specific phenomenon. Immunology 27: 457-462

37. Loker ES, Bayne CJ (1982) In vitro encounters between *Schistosoma mansoni* primary sporocysts and hemolymph components of susceptible and resistant strains of *Biomphalaria glabrata*. Am J Trop Med Hyg 31: 999-1005
38. Mattes MJ, Steiner LA (1978) Antisera to frog immunoglobulins cross-react with a periodate-sensitive cell surface determinant. Nature 273: 761-763
39. McKay D, Jenkins CR (1970) Immunity in the invertebrates. The role of serum factors in phagocytosis of erythrocytes by haemocytes of the freshwater crayfish (*Parachaeraps bicarinatus*). Aust J Exp Biol Med Sci 48: 139-150
40. Meuer SC, Cooper DA, Hodgdon JC, Jussey RE, Fitzgerald KA, Schlossman SF, Reinherz EL (1983) Identification of the receptor for antigen and major histocompatibility complex on human inducer T lymphocytes. Science 222: 1239-1242
41. Nappi AJ. Silvers M (1984) Cell surface changes associated with cellular immune reaction in *Drosophila*. Science 225: 1166-1168
42. Oss CJ van, Gillman CG (1972) Phagocytosis as a surface phenomenon I. Contact angles and phagocytosis on non-opsonized bacteria. J Reticuloendothel Soc 12: 283-289
43. Pahwa R, Ikehara S, Pahwa SG, Good RA (1979) Thymic function in man. Thymus 1: 27-58
44. Peterson PA, Kampe O (1981) Are the classic transplantation antigens primitive cell-surface antibodies? In: Reisfeld RA, Ferrone S (eds) Current trends in histocompatibility, vol 1. Plenum, New York, pp 437-455
45. Ratcliffe NA, Leonard C, Rowley AF (1984) Prophenoloxidase activation: Nonself recognition and cell cooperation in insect immunity. Science 226: 557-559
46. Ratcliffe NA, Rowley AF (eds) (1981) Invertebrate blood cells, vols 1 and 2. Academic Press London
47. Renwrantz LR, Cheng TC (1977) Identification of agglutinin receptors on hemocytes of *Helix pomatia*. J Invertebr Pathol 29: 88-96
48. Renwrantz LR, Cheng TC (1977) Agglutinin-mediated attachment of erythrocytes to hemocytes of *Helix pomatia*. J Invertebr Pathol 29: 97-100
49. Renwrantz LR, Stahmer A (1983) Opsonizing properties of an isolated hemolymph agglutinin and demonstration of lectin-like recognition molecules at the surface of hemocytes from *Mytilus edulis*. J Comp Physiol 149: 535-546
50. Roch P, Cooper EL, Eskinazi DP (1983) Serological evidence for a membrane structure related to human B_2-microglobulin expressed by certain earthworm leukocytes. Eur J Immunol 13: 1037-1042
51. Roch P, Valembois P (1978) Evidence for concanavalin A-receptors and their redistribution on lumbricid leukocytes. Dev Comp Immunol 2: 51-63
52. Roch P, Valembois P, DuPasquier L (1975) Response of earthworm leukocytes to concanavalin A and transplantation antigens. Adv Exp Med Biol 64: 45-54
53. Rowley AF, Ratcliffe NA (1980) Insect erythrocyte agglutinins. In vitro opsonization experiments with *Clitumnus extradentatus* and *Periplaneta americana* haemocytes. Immunology 40: 483-492
54. Sarmay G, Istvan L, Gergely J (1978) Shedding and reappearance of Fc, C3 and SRBC receptors on peripheral lymphocytes from normal donors and chronic lymphatic leukaemia (CLL) patients. Immunology 34: 315-321
55. Schlumpberger JM, Weissman IL, Scofield VL (1984) Separation and labeling of specific subpopulations of *Botryllus* blood cells. J Exp Zool 229:401-411
56. Schoenberg DA. Cheng TC (1980) Lectin-binding specificities of hemocytes from two strains of *Biomphalaria glabrata* as demonstrated by microhemadsorption assays. Dev Comp Immunol 4: 617-628
57. Schoenberg DA, Cheng TC (1981) Lectin-binding specificities of *Bulinus truncatus* hemocytes as demonstrated by microhemadsorption. Dev Comp Immunol 5: 145-149
58. Schoenberg DA, Cheng TC (1982) Concanavalin A-mediated phagocytosis of yeast by *Biomphalaria glabrata* hemocytes in vitro: Effects of temperature and lectin concentration. J Invertebr Pathol 39: 314-322
59. Shalev A, Greenberg AH, Logdberg L, Bjorck L (1981) B_2-microglobulin-like molecules in low vertebrates and invertebrates. J Immunol 127: 1186-1191

60. Sminia T, Knaap WPW van der, Edelenbosch P (1979) The role of serum factors in phagocytosis of foreign particles by blood cells of the freshwater snail, *Lymnaea stagnalis*. Dev Comp Immunol 3:37-44
61. Sminia T, Winsemius AA, Knaap WPW van der (1981) Recognition of foreignness by blood cells of the freshwater snail *Lymnaea stagnalis*, with special reference to the role and structure of the cell coat. J Invertebr Pathol 38: 175-183
62. Smith VJ, Ratcliffe NA (1978) Host defense reactions of the shore crab, *Carcinus maenas* (L), in vitro. J Mar Biol Assoc UK 58: 367-379
63. Smith VJ, Söderhäll K (1983) β1,3-glucan activation of crustacean haemocytes in vitro and in vivo. Biol Bull (Woods Hole) 164: 299-314
64. Spiegel E, Burger M, Spiegel M (1980) Fibronectin in the developing sea urchin embryo. J Cell Biol 87: 309-313
65. Stein E, Cooper EL (1981) The role of opsonins in phagocytosis by coelomocytes of the earthworm, *Lumbricus terrestris*. Dev Comp Immunol 5: 415-425
66. Stuart AE (1968) The reticulo-endothelial apparatus of the lesser octopus, *Eledone cirrosa*. J Pathol Bacteriol 96: 401-412
67. Toupin J, Lamoureux G (1976) Coelomocytes of earthworms: The T-cell-like rosette. Cell Immunol 26: 127-132
68. Toupin J, Lamoureux G (1976) Coelomocytes of earthworms: Phytohemagglutinin (PHA) responsiveness. In: Wright RK, Cooper EL (eds) Phylogeny of thymus and bone marrow-bursa cells. Elsevier/North Holland, Amsterdam, pp 19-25
69. Vasta GR, Cassels F (1983) N-acylaminosugar-specific lectins in the serum and a hemocyte microsomal fraction from the blue crab *Callinectes sapidus*. Am Zool 23: 901 (abstract)
70. Vasta GR, Cheng TC, Marchalonis JJ (1984) A lectin on the hemocyte membrane of the oyster (*Crassostrea virginica*). Cell Immunol 88: 475-488
71. Warr GW (1980) Membrane immunoglobulins of vertebrate lymphocytes. In: Marchalonis JJ, Cohen N (eds) Self/non-self discrimination. Plenum, New York, pp 141-170 (Contemporary Topics in Immunobiology, vol 9)
72. Williams AF, Gagnon J (1982) Neuronal cell Thy-1 glycoprotein: Homology with immunoglobulin. Science 216: 696-703
73. Yoshino TP (1981) Comparison of concanavalin A-reactive determinants on hemocytes of two *Biomphalaria glabrata* snail stocks: Receptor binding and redistribution. Dev Comp Immunol 5: 229-240
74. Yoshino TP (1982) Lectin-induced modulation of snail hemocyte surface determinants: Clearance of Con A-receptor complexes. Dev Comp Immunol 6: 451-461
75. Yoshino TP (1983) Surface antigens of *Biomphalaria glabrata* (Gastropoda) hemocytes: Occurrence of membrane-associated hemolymph-like factors antigenically related to snail hemoglobin. J Invertebr Pathol 41: 310-320
76. Yoshino TP (1983) Lectins and antibodies as molecular probes of molluscan hemocyte surface membranes. Dev Comp Immunol 7: 641-644
77. Yoshino TP, Granath WO Jr (1983) Identification of antigenically distinct hemocyte subpopulations in *Biomphalaria glabrata* (Gastropoda) using monoclonal antibodies to surface membrane markers. Cell Tissue Res 232: 553-564
78. Yoshino TP, Granath WO Jr (1985) Surface antigens of *Biomphalaria glabrata* (Gastropoda) hemocytes: Functional heterogeneity in cell subpopulations recognized by a monoclonal antibody. J Invert Pathol 45: 174-186
79. Yoshino TP, Renwrantz LR, Cheng TC (1979) Binding and redistribution of surface membrane receptors for concanavalin A on oyster hemocytes. J Exp Zool 207: 439-449

Note Added in Proof. To summarize this chapter, it is clear that our present knowledge concerning the structural and functional morphology of circulating blood cells, their surface membrane constituents, and their molecular interaction with soluble blood factors provides a complex picture of biochemical events which are only now beginning to be understood. Although there appears to be a general agreement that the classical vertebrate recognition factors such as surface Ig or T-cell antigen receptors are fundamentally distinct from those factors used by invertebrates in the recognition of foreignness, numerous examples of functional and structural analogies between blood cells of these two major animal groups hint at the possible existence of evolutionary ties in their immune systems. It is anticipated that further investigation of invertebrate blood cells and the molecules they express can only serve to strengthen our understanding of these phylogenetic relationships.

Chapter 3 Structure and Functions of Oyster Hemocytes

W. S. Fisher[1]

CONTENTS

1 Cell Morphology

There are a number of characteristics which should be considered in a definitive classification of hemocytes [40], but the initial criterion is most often morphology. Although no general agreement on the number of different cell types in oyster hemolymph has been reached, most investigators divide the hemocytes into at least two major classifications: the granular and the agranular [7,40]. There are apparent differences in the roles that granular and agranular cells play in any organism, and those roles are not necessarily the same for each species. The granular hemocytes are generally larger than the agranular cells and contain walled vesicles (granules) in the cytoplasm. Using light microscopy, investigators have also been able to distinguish acidophilic and basophilic granulocytes [7], stem cells [2], slightly granular cells [40], and differences based on nuclear size [28]. But even with ultrastructural investigation, these studies of oyster hemocytes have not led to a generally accepted scheme of nomenclature and classification. Rather, they have emphasized differences between oyster species and raised conflicting evidence within species. For example, there are major

1 Horn Point Environmental Laboratories, University of Maryland, P.O. Box 775, Cambridge, Maryland 21613, USA

Immunity in Invertebrates
Ed. by M. Brehélin

disagreements between investigators examining the ultrastructure of *Crassostrea virginica* agranular hemocytes [27,34,48]. Also, the easily distinguishable *C. virginica* granulocyte was first believed to be two cell types [30] due to its altered morphology upon degranulation [7].

Despite these conflicts, description of morphological types has been a major theme for published reports and new techniques for further differentiation are being explored [14,47]. Descriptive studies have been presented for *C. virginica* [14,27,30,34,47], *Ostrea edulis* [54], and *Crassostrea gigas* [2,49,50]. Although classifications of lamellibranch leukocytes have been postured since the turn of the century [40], no common scheme has evolved. This pursuit of morphological classification has opened the door for functional and biochemical studies, but does not invite meaningful reviews or comprehensive comparisons of hemocyte morphology [28]. Thus, structure and ultrastructure will be used in this report only as they specifically apply to functional concepts.

2 Functions of Oyster Hemocytes

2.1 Nutrition and Defense

Two major themes pursued by investigators have been the function of hemocytes in nutrition and defense [9,28]. The underlying reason for this dual role is the ability of hemocytes to recognize, locate, ingest, transport, and digest foreign particles. By the process of endocytosis (engulfment of fluids or particles into the cell), phagocytic hemocytes are able to move into the digestive tract to engulf particles of food and then migrate into the tissues to digest and release nutrient to the hemolymph and tissue cells [9]. They are also capable of reversing this process and capturing unwanted material in the hemolymph (to prevent contact or growth in host tissues) and transporting the material to the digestive tract for disposal in the feces. In this defensive role, the hemocytes may release captured particles via the digestive diverticula, alimentary tract, palps, pericardial cavity, or mantle cavity [52]. The epithelial tissue of the gonoducts, nephridia, and shell-forming mantle are not used as routes. Or, if the captured foreign material is digestible, it may be degraded intracellularly with a variety of enzymes stored in vesicles and granules. It has been postulated [7,28] that in evolution, phagocytosis was originally a nutrient-acquiring process and only later became associated with defense against invading organisms and foreign particles. The nutritive role has been reviewed by Takatsuki [54], Yonge [62], and Feng et al. [28]. A review of defensive responses to specific etiologic agents in bivalves is provided by Bayne [5] in a comprehensive report on molluscan defense systems. It is interesting that two oyster diseases are known to infect hemocytes; *Perkinsus marinum* in *C. virginica* [38, 43] and *Bonamia ostreae* in *O. edulis* [3,44].

Endocytosis is the general term for phagocytosis (cell-eating) and pinocytosis (cell-drinking), both of which are known to occur for oyster hemocytes. Although it has been shown [31] that all types of *C. virginica* hemocytes are phagocytic, it is clear that the granular cells are much more important in this respect [7]. Auffret [2] has shown ultrastructurally that small "lysosome-like" vesicles occur in *O. edulis* agranular cells, implying that they are potential phagocytes. Ruddell [49,50] observed that agranular cells were more actively phagocytic for *C. gigas*, but this has more recently been contested [28]. Regardless of the cell type, endocytosis is probably the single most important function of oyster hemocytes. A wide range of microorganisms and inorganic particles can be phagocytosed and pinocytosed by oyster hemocytes, including bacteria [4,11,25,31], yeast [20], viruses [1,22,33], red blood cells [26,55], india ink [52], and proteins such as bovine serum albumin, diptheria antitoxin, and gamma-globulin [23]. Hemocytes can remove 90% bacteria within 24 h and 99% in 72 h when exposed in vitro [58]. Phagocytosis is enhanced when the foreign materials are preincubated in hemolymph serum which contains an opsonic factor [57]. Energy for phagocytosis appears to be derived from a glycolytic pathway since hemocytes of the hard clam, *Mercenaria mercenaria*, utilize glucose and glycogen, produce lactate, and do not increase oxygen consumption during phagocytic activity [8].

2.2 Inflammation

Inflammation is the immediate response of tissue to a wound, which includes the constriction of blood vessels and sinuses as well as leukocytosis, or the infiltration of hemocytes to the site of the wound [42,49]. Stauber [53] informally defined leukocytosis as one of the cellular defensive activities of oysters (in addition to phagocytosis and encapsulation). More likely, leukocytosis is a mobilization for cellular defense, increasing the number of leukocytes available for the active defense processes of phagocytosis and encapsulation [15]. Hemocyte response to wounds and foreign invasions may vary with the species and the type of invader. Although granular cells are considered the most responsive cell for many bacteria and viruses in *C. virginica*, Farley [21] has shown that infection by the protozoan parasite *Haplosporidium nelsoni* elicits a response comprised almost solely of agranular cells.

2.3 Wound Repair

Hemocyte presence in large numbers at the site of a wound causes an "amebocyte plug" to form which restricts hemolymph loss and contamination. A wall of hemocytes surround the injury until they are replaced by epithelial cells growing over the plug. During this time, there is also a mass migration of hemocytes across the digestive epithelium to dispose of the necrotic and foreign material previously phagocytosed at the wound [42]. The hemocytes that remain at the wound may differentiate into fibroblast-like cells with intracytoplasmic fibrils. Fibroblast-like cells are occasionally found as circulating hemocytes [28], but are normally associated with wound repair sites [49] or at sites of encapsulation [48]. Hemocyte clotting in oyster encapsulation and wound repair is different than clotting in vertebrates. There is no fibrin or any other plasma involvement and the fusion between hemocytes is not permanent. That is, they can reverse the process and dissolve the clot by returning to an individual existence [40].

2.4 Encapsulation

Encapsulation occurs when groups of hemocytes surround a particle too large for phagocytosis. Cheng and Rifkin [15] have provided a detailed review of encapsulation and have attempted to classify the different capsules formed in terms of both the parasite and the host material used in the capsule. In many cases, there may be no reaction to a parasite due to coevolution of the host and parasite. The first host reaction may, in fact, be a response to degenerating host tissues. Leukocytes infiltrate the area and are believed to form fibroblast-like cells which deposit fibers in the intercellular matrix. The parasite is usually lined by a thin capsule of fibroblast-like cells which is then concentrically ringed by a very thick, fibrous and biochemically different capsule. Hemocytes are not preconditioned to either encapsulate or phagocytose, as evidenced by the fact that the first hemocytes that reach a parasite appear to attempt phagocytosis. Being incapable, due to the large size of the parasite, the cells stretch out into fibroblast-like cells and form the inner layer of the capsule. Postmortem studies have described hemocytes that were joined together as multinucleate giant cells [51].

2.5 Shell Repair

Bivalve leukocytes have also been reported to play a limited role in shell repair. For most species studied, the leukocytes migrate onto the mantle surface at the earliest stages of repair [60,61]. Although many researchers have reported that the leukocytes transport calcium to the site of shell repair, most calcium appears to come from other sources in the body [61]. It is possible that the primary activity of the leukocytes on the mantle epithelium is to protect the damaged epithelial layer during its repair process.

3 Hemocyte Activities and Their Structural Elements

The number of blood cells in lamellibranchs is extremely small in comparison to vertebrates, and circulating counts vary appreciably between individuals and over time [40]. This appears to hold true for oysters as well. Granular hemocytes observed in the hemolymph of *C. virginica* are round cells that move freely and independently in the vessels and sinuses [4]. Only circulating hemocytes are removed when hemolymph is drawn from the heart or adductor muscle, and these circulating cells do not represent the total population, which also includes mobile and "fixed" hemocytes in the tissues. Feng [24] has shown that the number of circulating hemocytes is dependent primarily on the temperature. This was believed to be a result of increased heart activity which agitated the cells in the vessels and sinuses so they would not settle.

3.1 Aggregation

When hemolymph is drawn into a syringe, the cells tend to aggregate into clumps [29, 30]. Narain [40] suggested that hemocyte clumping in lamellibranchs serves the same function as clotting in vertebrates, although there is no humoral fibrin in invertebrates. It has been shown that horseshoe crab (*Limulus*) leukocytes are solely responsible for stemming blood loss by their aggregation and contraction [37] and sea star (*Asterias*) leukocytes are hemostatic [36]. Aggregation of hemocytes presumably does not occur unless they are stimulated. Generally, contact and friction with foreign objects or mechanical agitation, such as withdrawal into a syringe, will serve as a stimulus. It is not known whether aggregation is the result of random collisions or whether hemocytes are chemically drawn toward one another. Takatsuki [54] observed in *O. edulis* that clumping hemocytes were held together by "bristle-like" processes. Later referred to as "spinous processes" [19], they are now called "filopodia" by Narain [40], who clearly outlined the morphologic and functional distinction between filopodia and pseudopodia. Filopodia are protoplasmic extensions which form interconnecting bridges that shorten and thicken to bring the hemocytes close together. Filopodia can be observed in EDTA-treated hemocytes of *C. virginica* and EDTA does not inhibit their aggregation [29]. However, leukocyte aggregation in several marine invertebrates has been found to be inhibited by chelating agents, such as EDTA and EGTA [36,37,41].

3.2 Transformation

Oyster hemocyte aggregates are not like mammalian clots because they are reversible. Aggregates of hemocytes from *C. virginica* that come into contact with a foreign surface, such as a glass slide, exhibit a characteristic pattern of exomigration by the peripheral cells (Fig. 1). But before the hemocytes move away from the clump, they must transform from their spherical to an ameboid shape. Under normal conditions, this hemocyte spreading process is complete within 10–15 min and the cells then become mobile. Spreading can be inhibited by the use of chelating agents or calcium- and magnesium-free water [29]. As long as calcium and magnesium are not added to the medium containing the hemocytes, they will remain spherical. It has also been found that an acute rise in salinity will retard the ability of *C. virginica* hemocytes to spread [29]. Apparently, the hemocytes need to be isoosmotic or slightly hyperosmotic to their environment in order to spread, and the amount of time required to spread increases as the acute increase in salinity becomes greater. The above results indicate a close link between cell spreading, osmotic adjustment, and, by association, cell volume regulation. This also implies that spreading is a cell membrane-related phenomenon. Obviously, there is relaxation of the rigid spherical structure, but there is also the cell machinery for volume regulation that resides in the plasmalemma [45,46]. Recent unpublished results show that spreading is also temperature-dependent with a Q_{10} of about 2. Chronic high temperature may increase cell membrane permeability and diminish its ability to spread.

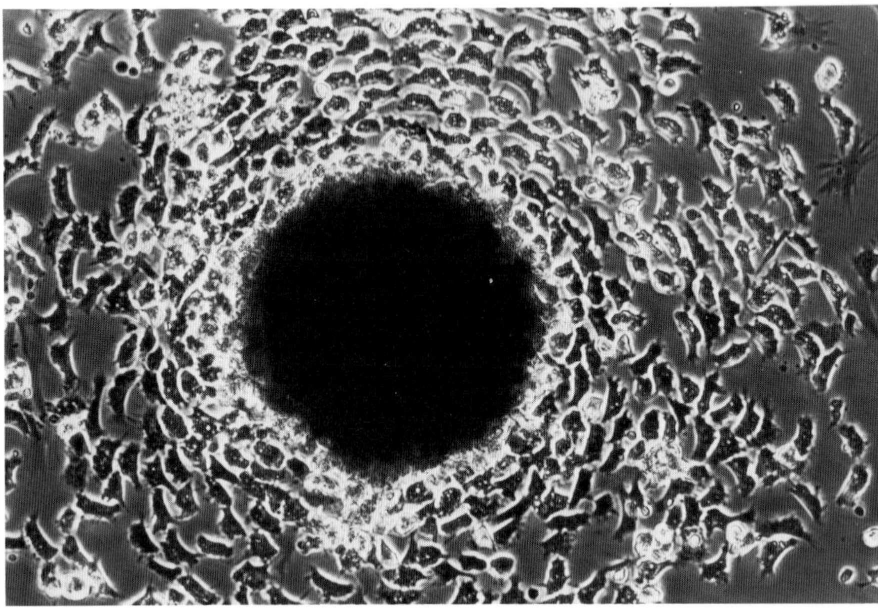

Fig. 1. Individual *C. virginica* hemocytes on the periphery of a large hemocyte aggregate transform to an ameboid shape and exomigrate. The crescent-shaped cell with the convex portion at the front is characteristic of locomotory *C. virginica* hemocytes

3.3 Locomotion

Cell locomotion does not occur until after the hemocytes are transformed to an ameboid shape. Basically, locomotion is the extension of a series of pseudopodia followed by the spread of ectoplasm and then endoplasm into the spaces between the pseudopods [4]. Pseudopodia are broad lamellate processes involved in cell movement and differ from filopodia which are more delicate and integral to clumping. A bundle of 290 microtubules has been observed along the length of a pseudopodium [7] and it appears that microtubules and microfilaments are necessary for locomotion since both colchicine and cytochalasin B inhibit random migration of *C. virginica* hemocytes [13]. Locomotion, like cell spreading, is affected by an acute rise in salinity. The speed of hemocyte movement is reduced when they are placed in salinities higher than their acclimation salinity [29]. Oysters from low salinity populations do not acclimate well to higher salinities since a month of acclimation to high salinities did not increase their rate of locomotion. Like spreading, the rate of locomotion is dependent on temperature with a similar Q_{10}. The mechanism of locomotion may well be the continuous invagination, or endocytosis, of the cell membrane which is then substituted at the leading edge of the moving hemocyte [6].

3.4 Chemotaxis

The direction of phagocyte movement is, of course, very important in locating and capturing foreign particles. However, studies on chemotaxis of oyster hemocytes are few. Cheng and Howland [12] have demonstrated chemotaxis toward the bacteria *Bacillus megaterium, Micrococcus varians*, and *Escherichia coli*, and have determined that the attractant was a chemical found in the cell walls and cell envelopes [35]. Hemocytes were not attracted to *Vibrio parahemolyticus*. Pretreatment of the *C. virginica* phagocytes with cytochalasin B and colchicine inhibited association with these bacteria, which implies that an intact cytoskeletal system (that is, both microfilaments and microtubules) are necessary for chemotactic response. However, these chemicals also had an effect on random migration [13].

3.5 Adherence

Bang [4] first described the "porcupine" appearance of bacteria adhered to a *C. virginica* hemocyte. He also showed that the hemocytes were selective and did not attach all bacteria. *Bacillus megaterium, Staphylococcus aureus,* and *Escherichia coli* are known to adhere to *C. virginica* hemocytes, and they adhere more readily at higher temperatures [31]. A possible mechanism for adherence was suggested by Vasta et al. [59]. They demonstrated a very high agglutination capacity from the microsomal fraction of hemolymph preparations. This implied that an agglutinin (lectin) was structurally bound in the cell membrane as a true cell-surface receptor. Thus, agglutination, normally perceived as a humoral component of the defense system, may also play a role in the adherence of foreign materials to phagocytes. Cheng [10] has briefly reviewed the growing number of invertebrate hemocytes found to have specific lectin binding sites on their surfaces.

3.6 Phagocytosis

Phagocytosis can be divided into at least four distinct stages; attraction, adherence, internalization (endocytosis), and intracellular digestion of foreign particulate matter. As mentioned earlier, diapedesis and exocytosis of the undigested particle into an excretory pathway may replace intracellular digestion. Phagocytosis of bacteria in oysters was first described by Bang [4] using a mixed culture of marine bacteria. *C. virginica* are known to phagocytose *Bacillus megaterium, Escherichia coli*, and *Staphylococcus aureus* [31]. Granular cells are clearly the most phagocytic for *C. virginica*, whereas both cell types are phagocytic for *C. gigas* [28]. In vitro studies on phagocytosis [31] have shown that increased temperatures enhance phagocytosis, supporting earlier studies using in vivo clearance rates [26]. Preliminary evidence from on-going research implies that the influence of temperature may be, at least partially, due to its effect on hemocyte locomotion which has a Q_{10} of about 2. At low temperatures, phagocytosis occurs, but there is a considerably lower digestive and migratory activity by hemocytes [26].

Internalization, or endocytosis, of foreign particles has been studied for a variety of vertebrate and invertebrate cell types. In oysters, a cell-membrane invagination at the site of foreign particle adherence results in an endocytotic vesicle with a unit membrane [7,27]. This vesicle, called a "primary phagosome", entraps the particle in the cytoplasm. The primary phagosomes may then fuse with granules which have a complex wall and are called "secondary phagosomes" [7,28]. B-glucuronidase, acid phosphatase, alkaline phosphatase, lipase, aminopeptidase, and lysozyme are present in hemocyte granules of *C. virginica* [16] which have been shown to be true lysosomes [63]. Peroxidase, nonspecific esterase, and acid phosphotases have been noted for granular hemocytes of *O. edulis* and *C. gigas* [2]. A hypersynthesis of these lysosomal enzymes was reported for molluscan phagocytes soon after they were challenged with bacteria [10]. When bacteria are destroyed intracellulary, glycogen granules are formed in the hemocytes, presumably for release to the hemolymph or for use by the hemocyte [11,18]. In so doing, hemocytes fulfill both their defensive and the nutritional roles at once.

Bacteria may also be destroyed by cellular products in the extracellular milieu. Tripp [56] showed that cell-free *C. virginica* hemolymph was cidal to certain bacteria and later detected lysozyme [39]. Lysozyme is believed to (1) destroy infectious agents such as bacteria; (2) alter foreign particles that have escaped detection so that they may be recognized as non-self; and (3) induce further inflammation at wound sites through autolytic action [32]. Release of lysosomal enzymes is a consequence of hemocyte "degranulation" where granules exocytose and spill their contents into the hemolymph plasma. Lysozyme was shown to decrease in hemocyte fractions and increase in serum fractions when *Mercenaria mercenaria* were exposed to the bacterium *Bacillus megaterium* [17]. Degranulation is an exocytotic process, but does not involve the rupture of the plasmalemma. Future research will undoubtedly focus on these interactions with the extracellular milieu, the humoral response, cell receptors, and the various stimuli for distinct hemocyte functions.

References

1. Acton RT, Evans EE (1968) Bacteriophage clearance in the oyster (*Crassostrea virginica*). J Bacteriol 95: 1260-1266
2. Auffret M (1984) Contribution of hematology to the knowledge of cellular immunity in bivalve molluscs (abstract). Invertebr Immunol Conf, Int Soc Dev Comp Immunol, Montpellier, France. Dev Comp Immunol (in press)
3. Bachere E, Comps M (1984) In: Pichot et al. 1980. Experimental infection of the flat oyster *Ostrea edulis* by the protistan *Bonamia ostreae* (abstract) 1st Int Colloq Pathol Marine Aquac, Montpellier, France, pp 37-38
4. Bang FB (1961) Reaction to injury in the oyster (*Crassostrea virginica*). Biol Bull (Woods Hole) 121: 57-68
5. Bayne CJ (1983) Molluscan immunobiology. In: Salleuddin ASM, Wilbur KM (eds) The Mollusca. Physiology, Part 2, Vol 4. Academic Press, New York, pp 407-486
6. Bretscher MS (1984) Endocytosis: relation to capping and cell locomotion. Science 224: 681-686
7. Cheng TC (1975) Functional morphology and biochemistry of molluscan phagocytes. Ann NY Acad Sci 266: 343-379

8. Cheng TC (1976) Aspects of substrate utilization and energy requirement during molluscan phagocytosis. J Invertebr Pathol 27: 263-268
9. Cheng TC (1977) Biochemical and ultrastructural evidence for the double role of phagocytosis in molluscs: defence and nutrition. In: Bulla LA Jr, Cheng TC (eds) Comparative Pathobiology, Vol 3. Invertebrate Immune Responses. Plenum, London, pp 21-30
10. Cheng TC (1983) The role of lysosomes in molluscan inflammation. Am Zool 23: 129-144
11. Cheng TC, Cali A (1974) An electron microscope study of the fate of bacteria phagocytized by granulocytes of *Crassostrea virginica*. Contemp Top Immunobiol 4: 25-35
12. Cheng TC, Howland KH (1979) Chemotactic attraction between hemocytes of the oyster *Crassostrea virginica*, and bacteria. J Invertebr Pathol 33: 204-210
13. Cheng TC, Howland KH (1982) Effects of colchicine and cytochalasin B on chemotaxis of oyster (*Crassostrea virginica*) hemocytes. J Invertebr Pathol 40: 150-152
14. Cheng TC Huang JW, Karadogan H, Renwrantz LR, Yoshino TP (1980) Separation of oyster hemocytes by density gradient centrifugation and identification of their surface receptors. J Invertebr Pathol 36: 35-40
15. Cheng TC, Rifkin E (1970) Cellular reactions in marine molluscs in response to helminth parasitism. In: Snieszko SF (ed) A Symposium on Diseases of Fishes and Shellfishes. Am Fish Soc (Wash DC), pp 443-496
16. Cheng TC, Rodrick GE (1975) Lysosomal and other enzymes in the hemolymph of *Crassostrea virginica* and *Mercenaria mercenaria*. Comp Biochem Physiol 52B: 443-447
17. Cheng TC, Rodrick GE, Foley DA, Koehler SA (1975) Release of lysozyme from hemolymph cells of *Mercenaria mercenaria* during phagocytosis. J Invertebr Pathol 25: 261-265
18. Cheng TC, Rudo BM (1976) Distribution of glycogen resulting from degradation of ^{14}C-labelled bacteria in the American oyster, *Crassostrea virginica*. J Invertebr Pathol 27: 259-262
19. Dundee DS (1953) Formed elements of the blood of certain fresh-water mussels. Trans Am Microsc Soc 72: 254-264
20. Eble AF, Tripp MR (1969) Oyster leucocytes in tissue culture: a functional study. Proc Natl Shellfish Assoc 59: 5
21. Farley CA (1968) *Minchinia nelsoni* (Haplosporidia) disease syndrome in the American oyster *Crassostrea virginica*. J Protozool 15: 585-599
22. Feng JS (1966) The fate of a virus, *Staphylococcus aureus* Phage 80, injected into the oyster, *Crassostrea virginica*. J Invertebr Pathol 8: 496-504
23. Feng SY (1965a) Pinocytosis of proteins by oyster leucocytes. Biol Bull (Woods Hole) 129: 95-105
24. Feng SY (1965b) Heart rate and leucocyte circulation in *Crassostrea virginica* (Gmelin). Biol Bull (Woods Hole) 128: 198-210
25. Feng SY (1966) Experimental bacterial infections in the oyster *Crassostrea virginica*. J Invertebr Pathol 8: 505-511
26. Feng SY, Feng JS (1974) The effect of temperature on cellular reactions of *Crassostrea virginica* to the injection of avian erythrocytes. J Invertebr Pathol 23: 22-37
27. Feng SY, Feng JS, Burke CN, Khairallah LH (1971) Light and electron microscopy of the leucocytes of *Crassostrea virginica* (Mollusca: Pelecypoda). Z Zellforsch Mikrosk Anat 120: 222-245
28. Feng SY, Feng JS, Yamasu T (1977) Roles of *Mytilus coruscus* and *Crassostrea gigas* blood cells in defense and nutrition. In: Bulla LA Jr, Cheng TC (eds) Comparative Pathobiology, Vol 3. Invertebrate Immune Responses. Plenum, London, pp 31-67
29. Fisher WS (1984) Osmotic and ionic effects on hemocyte function in the American oyster (abstract). Invertebr Immunol Conf, Int Soc Dev Comp Immunol, Montpellier, France, p. 70. Dev Comp Immunol (in press)
30. Foley DA, Cheng TC (1972) Interaction of molluscs and foreign substances: the morphology and behavior of hemolymph cells of the American oyster, *Crassostrea virginica*, in vitro. J Invertebr Pathol 19: 383-394
31. Foley DA, Cheng TC (1975) A quantitative study of phagocytosis by hemolymph cells of the Pelecypods *Crassostrea virginica* and *Mercenaria mercenaria*. J Invertebr Pathol 25: 189-197

32. Foley DA, Cheng TC (1977) Degranulation and other changes of molluscan granulocytes associated with phagocytosis. J Invertebr Pathol 29: 321-325
33. Fries CR, Tripp MR (1970) Uptake of viral particles by oyster leucocytes in vitro. J Invertebr Pathol 15: 136-137
34. Hawkins WE, Howse HD (1982) Ultrastructure of cardiac hemocytes and related cells in the oyster *Crassostrea virginica.* Trans Am Microsc Soc 101 (3): 241-252
35. Howland KH, Cheng TC (1982) Identification of bacterial chemoattractants for oyster (*Crassostrea virginica*) hemocytes. J Invertebr Pathol 39: 123-132
36. Kanungo K (1982) In vitro studies on the effects of cell-free coelomic fluid, calcium, and/or magnesium on clumping of coelomocytes of the sea star *Asteria forbesi* (Echinodermata: Asteroidea). Biol Bull (Woods Hole) 163:438-452
37. Kenney DM, Belamarich FA, Shepro D (1972) Aggregation of horseshoe crab (*Limulus polyphemus*) amebocytes and reversible inhibition of aggregation by EDTA. Biol Bull (Woods Hole) 143: 548-567
38. Mackin JG, Owen HM, Collier A (1950) Preliminary note on the occurence of a new protistan parasite, *Dermocystidium marinum* n.sp. in *Crassostrea virginica* (Gmelin). Science 111: 328-329
39. McDade JE, Tripp MR (1967) Lysozyme in the hemolymph of the oyster, *Crassostrea virginica*. J Invertebr Pathol 9: 531-535
40. Narain AS (1973) The amoebocytes of lamellibranch molluscs, with special reference to the circulating amoebocytes. Malacol Rev 6: 1-12
41. Noble PB (1970) Coelomocyte aggregation in *Cucumaria frondosa*: Effect of ethylenediaminetetraacetate, adenosine, and adenosine nucleotides. Biol Bull (Woods Hole) 139: 549-556
42. Pauley GB, Sparks AK (1965) Preliminary observations on the acute inflammatory reaction in the Pacific oyster, *Crassostrea gigas* (Thunberg). J Invertebr Pathol 7: 248-256
43. Perkins FO, Menzel RW (1967) Ultrastructure of sporulation in the oyster pathogen *Dermocystidium marinum*. J Invertebr Pathol 9: 205-229
44. Pichot Y, Comps M, Tige G, Grizel H, Rabouin MA (1979) Reserches sur *Bonamia ostreae* gen. n., sp. n., parasite nouveau de l'huitre plate *Ostrea edulis* (L). Rev Trav Inst Peches Marit 43: 131-140
45. Pierce SK (1982) Invertebrate cell volume control mechanisms: A coordinated use of intracellular amino acids and inorganic ions as osmotic solute. Biol Bull (Woods Hole) 163: 405-419
46. Pierce SK, Amende LM (1981) Control mechanisms of amino acid-mediated cell volume regulation in salinity-stressed molluscs. J Exp Zool 215: 247-257
47. Renwrantz LR, Yoshino TP, Cheng TC, Auld KR (1979) Size determination of hemocytes from the American oyster, *Crassostrea virginica*, and the description of a phagocytosis mechanism. Jahrb Zool Abt Physiol Zoomorph 83: 1-12
48. Rifkin E, Cheng TC, Hohl HR (1969) An electron microscope study of the constituents of encapsulating cysts in *Crassostrea virginica* formed in response to *Tyolcephalum metacestodes*. J Invertebr Pathol 14: 211-226
49. Ruddell CL (1971a) The fine structure of oyster agranular amebocytes from regenerating mantle wounds in the Pacific oyster, *Crassostrea gigas*. J Invertebr Pathol 18: 260-268
50. Ruddell CL (1971b) The fine structure of the granular amebocytes of the Pacific oyster, *Crassostrea gigas*. J Invertebr Pathol 18: 269-275
51. Sparks AK, Pauley GB (1964) Studies of the normal post-mortem changes in the oyster, *Crassostrea gigas* (Thunberg). J Insect Pathol 6: 78-101
52. Stauber LA (1950) The fate of India ink injected intracardially into the oyster, *Ostrea virginica* (Gmelin). Biol Bull (Woods Hole) 98: 227-241
53. Stauber LA (1961) Immunity of invertebrates, with special reference to the oyster. Proc Natl Shellfish Assoc 50: 7-20
54. Takatsuki S (1934) On the nature and functions of the amoebocytes of *Ostrea edulis*. Q J Microsc Sci 76: 379-431
55. Tripp MR (1958) Disposal by the oyster of intracardially injected red blood cells of vertebrates. Proc Natl Shellfish Assoc 48: 143-147

56. Tripp MR (1960) Mechanisms of removal of infected microorganisms from the American oyster, *Crassostrea virginica* (Gmelin). Biol Bull (Woods Hole) 119: 210-223
57. Tripp MR (1966) Hemagglutinin in the blood of the oyster *Crassostrea virginica.* J Invertebr Pathol 8: 478-484
58. Tripp MR, Kent VE (1967) Studies on oyster cellular immunity. In Vitro 3: 129-135
59. Vasta GR, Sullivan JT, Cheng TC, Marchalonis JJ, Warr GW (1982) A cell membrane-associated lectin of the oyster hemocyte. J Invertebr Pathol 40: 367-377
60. Wagge LE (1951) The activity of amebocytes and of alkaline phosphatases during the regeneration of the shell in the snail *Helix aspersa.* Q J Microsc Sci 92: 3077-3210
61. Watanabe N (1983) Shell repair. In: Wilbur KM (ed) The Mollusca, Vol 4. Physiology, Part 1. Academic Press, New York, pp 289-316
62. Yonge CM (1946) Digestion of animals by lamellibranchs. Nature 157: 729
63. Yoshino TP, Cheng TC (1982) Experimentally induced elevation of aminopeptidase activity in hemolymph cells of the American oyster, *Crassostrea virginica.* J Invertebr Pathol 27: 267-270

Chapter 4 Insect Haemocytes: A New Classification to Rule Out the Controversy

M. Brehélin[1] and D. Zachary[2]

CONTENTS

1 Introduction

It has often been remarked that the classification of insect haemocytes is the subject of numerous controversies. This was seen once again at the round table discussion devoted to these cells at the Invertebrate Immunology Conference in Montpellier. Several explanations can be found for this. Some are connected with the methods used in research: method of preparation of the cells, lack of clear-cut tinctorial affinities in optical microscopy. Others are associated with the physiology of haemocytes: rapid transformation at bleeding, possibility of differentiation in circulating blood. Finally, another category of possible explanations concerns the scientists themselves: the terminology used is not always free of ambiguity, and it is difficult to make analogies between cell types described using electron microscopy and those described using optical microscopy.

However, it would seem possible for haematologists to reach an agreement (1) if only reliable morphological characteristics of the ultrastructure are taken into account,

1 Laboratoire de Pathologie Comparée, ERPAM, Université des Sciences et Techniques du Languedoc, 34060 Montpellier, France

2 Laboratoire de Biologie Générale, Université Louis Pasteur, 12 rue de l'Université, 67000 Strasbourg, France

Immunity in Invertebrates
Ed. by M. Brehélin

(2) if fresh terminology is used for the cells which are the object of the controversy, and (3) if only haemocyte functions about which there is agreements are retained. We have attempted to draw up such an agreement in the present review.

We have described the ultrastructural features of each haemocyte which enable the definition of the type of cell. We have described their behaviour at bleeding and under in vitro conditions. Their in vivo functions are specified when these have been substantiated unquestionably. Two types of terminology have been used to name cell types. We have used the names found in scientific literature for those where there is complete agreement. The other types, which are the subject of controversy, have been designated by the term granular haemocyte (GH) followed by a number (GH1, GH2, etc).

2 Insect Haemocyte Types

Distinction has thus been made between nine types of insect haemocytes: prohaemocytes, plasmatocytes, oenocytoids, spherule cells, thrombocytoids, and four types of granular haemocyte.

2.1 Prohaemocytes (Pr)

They are small rounded cells with a high nuclear-cytoplasmic ratio, numerous free ribosomes, and a poorly developed rough endoplasmic reticulum (RER). They show no inclusions other than cytoplasmic organelles and little sign of a visible differentiation.

They often exhibit a high mitotic index and they are considered as stem cells of circulating haemocytes.

No transformations are visible under in vitro conditions.

They are present in all species studied to date. They are the "prohaemocytes" of the literature.

Remarks: Quantitative studies are needed to establish their functions as stem cells.

2.2 Plasmatocytes (Pl) (Fig. 1)

Under in vivo conditions or when fixed at bleeding, they are most often spindle-shaped. They sometimes have some vesicles at the periphery or a few extensions of the plasma membrane. The free ribosomes are numerous; their RER is generally well developed but always in narrow elongated cisternae never enlarged. They show no granular inclusions; sometimes a few primary lysosomes as well as rare multivesicular bodies are visible. These lysosomes are never "condensed" into electron-dense granules.

They exhibit a variable amount of phagocytosis which is very limited in vivo [1,6, 9,22]. They also would be a step in a process of differentiation from Pr to other hae-

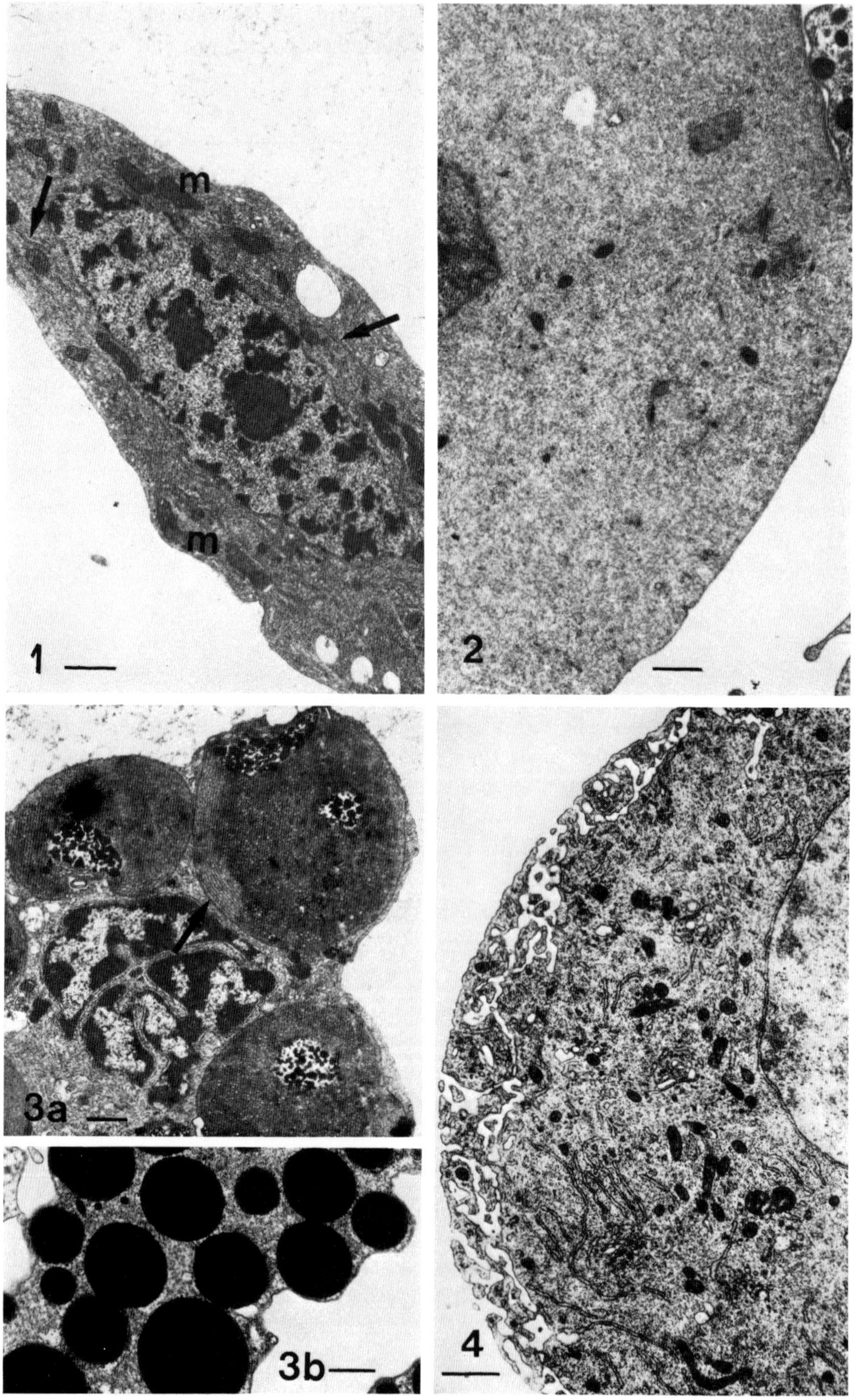
m
m
1
2
3a
3b
4

mocyte types. They form capsules and nodules around foreign bodies, in the species in which they are present (see Chap. 12).

No lysis of these cells is visible under in vitro conditions. But we can observe modifications in shape such as the formation of veils and spreading of the cell on the substratum.

They are present in all lepidopteran species studied to date and some species of Diptera.

In the literature they have been called "agranular plasmatocytes" [26], "typical plasmatocytes" [7] or "lamellocytes" [27] in *Drosophila* species (see Chap. 13).

2.3 **Oenocytoids (Oe)** (Fig. 2)

These are large cells with a low nuclear-cytoplasmic ratio. The nucleus is most often rounded with an eccentric location within the cell. The homogeneous cytoplasm exhibits numerous free ribosomes. The other typical cytoplasmic organelles (mitochondria, Golgi apparatus, RER), are very little developed. When present the rare cisternae of the RER are short and enlarged as the perinuclear cisterna. In most species, Oe are devoid of granular inclusions, but it is possible to observe, very occasionally some dense granules. In some species, especially lepidopterans, the cytoplasm of Oe contains numerous microtubules or fibers, or is filled with large crystal-like inclusions as in *Drosophila* species (see Chap. 13).

The cytoplasm of Oe possesses a phenol-oxidase activity which can be detected after incubation with tyrosine, DOPA or DOPAmine [11,13].

The functions of Oe are almost unknown, but the presence of the phenol-oxidase activity indicates that they could play a role in the metabolism of melanins and tanning agents, for instance when capsules are formed around foreign bodies or at the time of cuticle tanning. It has been shown by Rizki and Rizki [28] that the crystal observed in *Drosophila* species is tyrosine.

These cells remain generally stable under in vitro conditions, at bleeding, but rupture of Oe have been observed especially in *Drosophila*, and this, together with their phenol-oxidase activity may also indicate a role for these cells in non-self recognition (see Chaps. 12 and 15).

Oenocytoids seem to be a widely distributed type of haemocyte among insect species. But in each species they are rare in the circulating blood. They are absent in *Locusta migratoria* in which some haemocytes have been wrongly termed Oe by Hoffmann and co-workers [20]. They are the "oenocytoids" of the literature.

Fig. 1. Plasmatocyte: absence of peculiar inclusions; the rough endoplasmic reticulum (RER) is present in narrow and elongated cisternae (*arrows*). *M* = mitochondria. *Cirphis unipuncta* (Lepidoptera). Transmission electron microscopy (TEM)

Fig. 2. Oenocytoid: in this large cell the organelles are rare, in a uniformly dense cytoplasm. *Melolontha melolontha* (Coleoptera). (TEM)

Fig. 3a,b. Spherule cells: **a** In the Lepidoptera *C. unipuncta*, the spherules show an internal substructure (*arrow*). Note that the shape of this cell is imposed by the large and numerous inclusions. (TEM). **b** In the Coleoptera *M. melolontha*, the spherules are uniformly electron dense. (TEM)

Fig. 4. Thrombocytoid: Note the numerous and intricate invaginations of the plasma membrane which sometimes reach the vicinity of the nucleus. *Calliphora erythrocephala* (Diptera). *Bars* = 1 μm

Remarks: Several authors have shown pictures of haemocytes which seem to be transitional forms between granular haemocytes and Oe. This suggests a possible developmental relationship between these two types of haemocytes.

2.4 Spherule Cells (Sph) (Fig. 3a,b)

These large haemocytes have a shape resulting from the large and numerous inclusions that they contain. In the same way, their nucleus appears distorded by the inclusions. The RER is often well-developed. The typical inclusions of this haemocyte type are the "spherules". They are membrane bounded, at least three times larger than the granules of granular cells (see below). Most often they are regular in shape (rounded) and exhibit an internal substructure in some species. In Lepidoptera for instance, several types of spherule structure can be evidenced such as a crystal-like lattice in concentric layers in *Spodoptera littoralis* [18] or *Heliothis armigera* [14]. In other species, such as *Melolontha melolontha* [12] or *Clitumnus extradentatus* [33], the spherules are electron dense without any visible structure. All these spherules originate from Golgi complexes which are numerous in cells where spherules are few.

Their functions are actually unknown. Some authors [1,17] suggest that they can play a role in synthesis of mucopolysaccharides of the blood, and Nittono [23] reports that there is a correlation between the number of these haemocytes and the amount of silk synthesized by *Bombyx mori*. It is certain that the Gupta and Sutherland's hypothesis [17] that the spherules emitted from Sph transform into other haemocyte types in the haemolymph of cockroaches, has to be discarded.

In some cases, under in vitro conditions, the spherules can be released by rupture of the cells.

These haemocytes are widely distributed among insect species. Sometimes (e.g. in *M. melolontha*), they are numerous in the circulating blood. They seem to be absent in dipterans. They are the "spherule cells" of most authors, called "adipohemocytes" by Harpaz et al. [18] and "spherulocytes" by Gupta [16].

Remarks: Sph-like cells have been reported in the blood of *Calliphora* at the time of pupation, in studies conducted under light microscopy. They are in fact phagocytic haemocytes overloaded with engulfed tissue debris at metamorphosis [36] and are not true Sph.

2.5 Thrombocytoids (Th) (Fig. 4)

These haemocytes have a regular general shape. Their plasma membrane exhibits intricate invaginations which often can reach the vicinity of the nucleus. These cells sometimes disintegrate in the blood, leading to the formation of non-nucleated cytoplasmic fragments, each surrounded by an intact plasma membrane. Around the nucleus, only a narrow ring of cytoplasm remains visible: they are the naked nuclei of the literature. In *C. erythrocephala*, they have a RER characterized by long, parallel and systematically curved cisternae. In the same species they contain tiny dense granules.

In an other dipteran, *Tipula paludosa* [10] the numerous inclusions observed in the cytoplasm are larger than those of *Calliphora* and exhibit an internal structure.

Intact Th phagocytose bacteria under in vitro conditions in *C. erythrocephala* [30] but not under in vivo conditions [35]. Cytoplasmic fragments as well as entire Th form capsules around foreign bodies and seal off the wounds [37].

The disintegration of Th giving rise to cytoplasmic fragments, occurs in vivo, in the circulating blood; other special modifications of the cells do not occur in vitro.

These cells have been described only in dipteran species [21,37] but seem to be absent in *Drosophila* and in *Culicidae*. They have been termed "thrombocytoids" by Zachary and Hoffmann [36] due to their resemblance with the thrombocyte of vertebrates, and "granular cells" by Carter et al. [10]. As mentioned above the cytoplasmic fragments and naked nuclei observed by numerous authors under the light microscope, originate from this haemocyte type.

Remarks: Other uncommon haemocytes which possess very elongated protoplasmic expansions and small membrane-bounded dense inclusions in their cytoplasm have been observed in *M. melolontha* [12], *Holotrichia kiotoensis* [2], two coleopteran species, and in few Lepidoptera such as *Cirphis unipuncta* [5]. They have been termed "podocytes" and "vermiform cells". We tentatively include these haemocytes in Th, despite the fact that they never disintegrate in the blood and do not participate in the formation of capsules.

2.6 Granular Haemocytes (GH)

They have been divided into four subtypes, according to their ultrastructural characteristics, especially nature, sight and origin of their typical inclusions, called granules.

2.6.1 Granular Haemocytes 1 (GH1) (Figs. 5 and 6)

These can be round or oval cells sometimes with numerous cytoplasmic digitations and pinocytotic vesicles. Their RER is always very well developed, most often in enlarged cisternae, depending on the developmental stage. Numerous Golgi complexes are evident within the cell. They contain three types of granular inclusions. The first type are membrane-bound bodies with a microtubular substructure and they are formed from Golgi vesicles. The second type of inclusions are granules uniformly electron-dense and also membrane bounded. It is certain that at least some of the dense granules are structured ones in which the microtubular structure is no longer visible; this structure reappears when the granules are discharged from the cell. The last type of inclusions are heterogeneous bodies and multivesicular bodies, resorptive in nature. The inclusions of this last type are larger and more numerous at the end of the larval instars (time of metamorphosis).

Granular haemocytes 1 exhibit phagocytic capacities against living as well as "inert" (nonimmunogen) material present in the haemolymph at the time of tissue histolysis or after injection. They seem to be the most important haemocytes to react in the processes of foreign material recognition and haemolymph coagulation [8,24,33]. In

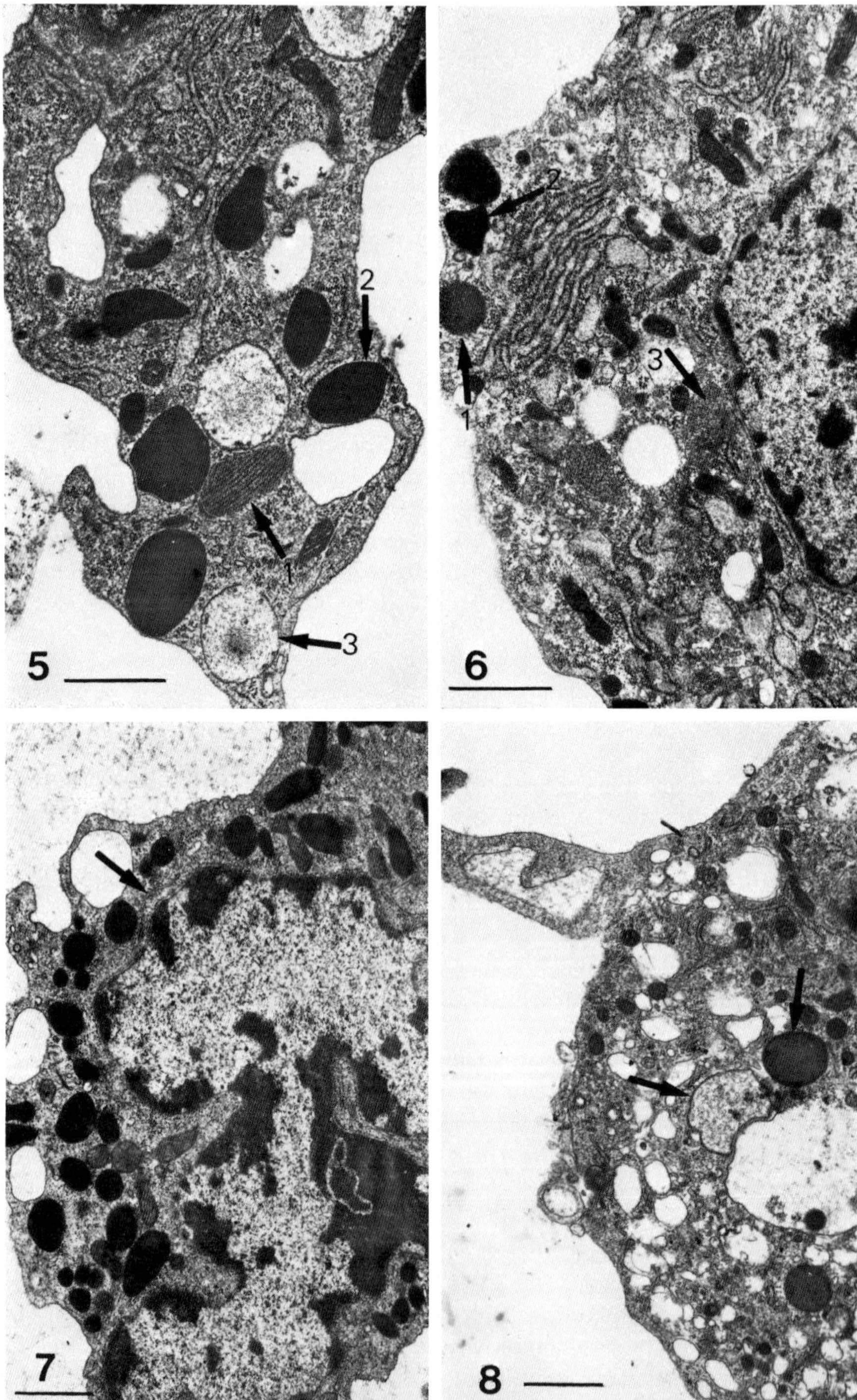
5
2
1
3
6
2
1
3
7
8

haemolymph, in contact with foreign material, they exhibit transformations such as the extrusion of some granules (for instance in lepidopteran species) or a complete lysis of the cell (for instance in *L. migratoria*). These transformations seem to release into the blood, some factors which are going to coat the foreign particles and make them clump, thus allowing their recognition by other cell types before endocytosis or encapsulation [5,8,24,26].

Under in vitro conditions these cells exhibit the same transformations as those described above. In culture conditions they are no longer visible after a few hours in vitro in *Galleria mellonella* [31,32]. In *L. migratoria*, we have observed two populations of GH1 [6], one with highly unstable cells and the other with cells which show slight or no modifications after 15 min in vitro. All the cells of these two populations exhibit the same ultrastructural features as well as the endocytotic capacities in vivo.

All the insect species studied to date possess GH1, except those belonging to the Diptera (or at least some of them: *Culicidae, Calliphoridae, Drosophilidae, Glossina*; perhaps they are present in *Tipula* larvae).

Some of the "granular cells" of Ratcliffe and co-workers [26], the "coagulocytes" of Hoffmann et al. [20], the "plasmatocytes" of Wigglesworth in *Rhodnius prolixus* [34], the "adipohaemocytes" of Beaulaton [4] and some of the "plasmatocytes" and "granulocytes" of Akai and Sato [2] are GH1.

Remarks: Depending on the stage of development in the same insect, the number of granules as well as the ratio between the three inclusion types may vary [14]. In the same way, the RER which is always well developed, sometimes appears as very numerous and enlarged cisternae filled with moderately dense material, depending on the age in the instar.

2.6.2 Granular Haemocytes 2 (GH2) (Fig. 7)

These haemocytes are polymorphic cells, sometimes regular in shape and sometimes exhibiting numerous digitations. Their RER is moderately or little developed, always in narrow cisternae, never enlarged. The Golgi complexes are generally numerous. The most typical features of this haemocyte type are the dense membrane-bounded granules, which fill the cytoplasm; they are of various sizes and shapes, but always electron-dense. These granules are synthesized by the Golgi apparatus, by fusion of small vesicles. In the "native form" (near the Golgi complex) they are sometimes less electron dense.

Fig. 5. Granular haemocyte 1: the three kinds of inclusions are visible in this cell: *1* = structured granules; *2* = uniformly dense granules; *3* = heterogeneous bodies of resorptive nature. Note the development of the RER. *Locusta migratoria* (Orthoptera) (TEM)

Fig. 6. Granular haemocyte 1: the same three kinds of inclusions are visible in this cell, as well as the important development of the RER. *Galleria mellonella* (Lepidoptera) (TEM)

Fig. 7. Granular haemocyte 2: only electron-dense granules are visible in this cell; the RER is present in narrow cisternae (*arrow*). *Periplaneta americana* (Dictyoptera) (TEM)

Fig. 8. Granular haemocyte 3: the inclusions (*arrows*) of this cell are only primary lysosomes and resorptive bodies. *L. migratoria* (TEM). *Bars* = 1 μm

They show little or no phagocytic capacities in vivo. In species in which they are present, they form the bulk of capsules around foreign bodies. For instance in *L. migratoria*, they encapsulate implanted foreign bodies too large to be phagocytosed by the phagocytic cells. In these capsules, GH2 undergo profound modifications such as extreme flattening of the cells, formation of desmosome-like junctions and synthesis of numerous microtubules in their cytoplasm. In the same species, they also act in nodule formation and wound healing.

Under in vitro conditions they seem to lose some of their granules.

They are present in only a few species, especially those belonging to the orthopteroid complex and in larvae of *M. melolontha* (Coleoptera). They are absent in most of the lepidopteran species studied to date and in the Diptera species.

They are the "granulocytes" of Hoffmann and Stoeckel [19], Brehélin and co-workers [8], some of the "granular plasmatocytes" of Ratcliffe and co-workers [25,26].

Remarks: In some species they show peculiarities such as concentric tubules in vacuoles of *Periplaneta americana* [3] or single tubules in the RER of *L. migratoria* [20]. They are not phagocytes overloaded with engulfed material because (1) they do not show endocytotic capacities, (2) the bulk of their inclusions are homogeneous, entirely synthesized by the Golgi apparatus and are not resorptive bodies.

2.6.3 Granular Haemocytes 3 (GH3) (Fig. 8)

These generally large cells exhibit numerous digitations and pinocytotic vesicles. Their RER is well developed, sometimes forming enlarged cisternae. The Golgi complexes are numerous as are the primary lysosomes. Heterogeneous bodies are sometimes few but most often numerous; these are auto- and (or) heterophagosomes whose content is sometimes very dense and make these haemocytes look like granular cells, at least at the light microscopic level. But under electron microscopy one can see the heterogeneity of these granules which are multivesicular or myelin-like bodies with a membranous or "fibrillar" substructure. They exhibit an acid phosphatase activity.

All these characteristics endow the GH3 with an important role in endocytosis. In fact these cells very actively engulf all the particles that we inject, the living ones (red blood cells, bacteria) as well as the inert (nonimmunogen) ones. These haemocytes are very active at the time of metamorphosis, especially in Diptera in which they engulf parts or the histolyzing tissues.

They usually remain stable under in vitro conditions and can spread on the substratum. Sometimes the overloading with engulfed material causes cell breakdown.

They are present in numerous species and especially in those of Orthoptera and Dictyoptera. They seem to be absent in Lepidoptera but are numerous in Diptera.

Remarks: When they become loaded with engulfed material, for instance at the time of metamorphosis, these cells look like other types of granular haemocytes (GH2) or spherule cells, with which they can be confused at the light microscopic level.

2.6.4 Granular Haemocytes 4 (GH4)

They are round or flattened cells with a high nuclear-cytoplasmic ratio. Their RER is little developed. The perinuclear cisterna is often swollen but this is most likely an artifact due to difficulties of fixation of these cells. Small numbers of polyribosomes are present. Their cytoplasm contains membrane-bounded pleomorphic granules which are electron-dense or sometimes show an internal organization in two distinct regions: an inner electron-dense core and an outer granular zone [29].

These cells are highly unstable in vitro and their rapid degranulation, together with the transformations observed in GH1 led to the coagulation of the plasma [29].

This haemocyte type seems to have been observed especially in Cheleutoptera: *Carausius morosus* [15] and *Clitumnus extradentatus* [29].

3 Discussion and Conclusions

The numerous studies carried out using optical microscopy have demonstrated their limits with regards to the obtaining of criteria for the definition of types of insect haemocyte. On the other hand, images obtained using electron microscopy show that features common to all species of insect studied to date can be revealed. This is why we decided to study the types of haemocyte and to define them according to the function of their ultrastructural characteristics.

Part of the controversy between haematologists is based on the analogies that must be made between work carried out using optical microscopes and electron microscopes. These difficulties can be avoided to a certain extent if the characteristics provided by the in vitro behaviour of haemocytes and the study of their functions are added to their morphological features. It is possible, for example, to separate GH1 from GH2: the latter remain stable at bleeding whereas most of the former are transformed rapidly. GH1 haemocytes play a role in the recognition of non-self [8,24,33] whereas GH2 cells form the bulk of capsules around foreign bodies [5]. Likewise, GH1 haemocytes possess a considerable endocytotic capacity while this reaction is limited in GH2.

Another point that should be stressed also concerns haemocyte functions. The same function can be fulfilled by cells of different types. This is the case for capsule formation. Capsules consist of GH2 in *L. migratoria* and *M. melolontha* whereas they are formed by Pl in lepidopterans. Likewise, none of these haemocytes possesses great phagocytic capacity in vivo. It is therefore tempting to consider that they belong to the same type of haemocyte. This may be the case, but the presence of dense granules in only some of the cells makes it necessary to make a distinction between them. At the present state of knowledge, there is nothing to confirm that they are different physiological states of the same cell. Moreover, even if they are the same type of haemocyte, the fact that they display different morphological features obliges us to assign a name to each. It is also possible to find functional analogies between various granular haemocytes, e.g. phagocytis by GH1 and GH3, or the inducing of blood clotting by GH1 and GH4. This is why we have not given the classic names used for haemocytes in publications since most of these denominations refer to functions: "coagulocyte", "macrophagic plasmatocytes", etc. The fact that we have given them the same initials followed by a number does not mean that they are different stages in the differentiation of a single type of haemocyte, or that they are different physiological states of the same cell. The terminology described here signifies that we agree on the definition of the morphological characteristics and, whenever possible, on the types of functions fulfilled, enabling a description of each type of haemocyte.

We propose the following procedure for research work on insect haemocytes:

1. Ultrastructural study of the cells fixed immediately after bleeding, and then of the other cells fixed a few minutes after bleeding. It should also be borne in mind that the definition of the haemocyte type should not be based on a single image but on views of different sections in order to be able to observe enough characteristic structures.
2. Determination of which cells are involved in various functions such as coagulation, phagocytosis, or capsule and nodule formation.
3. Study of several enzymatic activities such as acid phosphatase and phenol-oxidases.

We hope that the definitions given in this article will make it possible to end the controversy. Research work is in progress to find answers to two types of question: (a) What relationships exist between different types of haemocyte, i.e. do they belong to different cell lineages or are they different steps in the differentiation of a single lineage? (b) What exactly are the functions of each type, how do they carry out their functions, and what collaboration is possible between different cell types?

Acknowledgements. We are grateful to Prof. N.A. Ratcliffe for reading the manuscript and helpful suggestions.

References

1. Akai H, Sato S (1973) Ultrastructure of the larval haemocytes of the silkworm, *Bombyx mori* (L) (Lepidoptera, Bombycidae). Int J Insect Morphol Embryol 2: 207-231
2. Akai H, Sato S (1979) Surface and internal ultrastructure of hemocytes of some insects. In: Gupta AP (ed) Insect Hemocytes, development, forms, functions and techniques. Cambridge University Press, Cambridge, London, p 129
3. Baerwald RJ, Boush GM (1970) Fine structure of the haemocytes of *Periplaneta americana* with reference to marginal bundles. J Ultrastruct Res 31: 151-161
4. Beaulaton J (1968) Etude ultrastructurale et cytochimique des glandes prothoraciques des vers à soie aux quatrième et cinquième âges larvaires I. La *tunica propria* et ses relations avec les fibres conjonctives et les hémocytes. J Ultrastruct Res 23: 474-498
5. Brehélin M (1977) Etude morphologique et fonctionnelle des hémocytes d'insectes. Thesis, L. Pasteur University, Strasbourg
6. Brehélin M, Hoffmann JH (1980) Phagocytosis of inert particles in *Locusta migratoria* and *Galleria mellonella*: study of ultrastructure and clearance. J Insect Physiol 26: 103-111
7. Brehélin M, Zachary D (1983) About insect plasmatocytes and granular cells. Dev Comp Immunol 7: 683-686
8. Brehélin M, Hoffmann JA, Matz G, Porte A (1975) Encapsulation of implanted foreign bodies in *Locusta migratoria* and *Melolontha melolontha*. Cell Tissue Res 160: 283-289

9. Brehélin M, Zachary D, Hoffmann JA (1978) A comparative ultrastructural study of blood cells from nine insect orders. Cell Tissue Res 195: 45-57
10. Carter JB, Green EI, Barley P (1982) The haemocytes of *Tipula paludosa* larvae. Proc ISDCI Invertebr Immunol Conf, September 1982, Swansea
11. Crossley AC (1975) The cytophysiology of Insect Blood. Adv Insect Physiol 11: 117-222
12. Devauchelle G (1971) Etude ultrastructurale des hémocytes du coléoptere *Melolontha melolontha* (L). J Ultrastruct Res 34: 492-516
13. Drif L, Brehélin M (1983) The circulating hemocytes of *Culex pipiens* and *Aedes aegypti*: cytology, histochemistry, hemograms and functions. Dev Comp Immunol 7: 687-690
14. Essawy M, Maleville A, Brehélin M (1985) The haemocytes of *Heliothis armigera* (Lepidoptera): ultrastructure and modifications in the course of larval development. J Morphol (in press)
15. Goffinet G, Grégoire C (1975) Coagulocyte alterations in clotting haemolymph of *Carausius morosus* (L). Arch Int Physiol Biochim 83: 702-722
16. Gupta AP (1979) Hemocyte types: their structures, synonymies interrelationships and taxonomic significance. In: Gupta AP (ed) Insect Hemocytes: Development, forms, functions and techniques. Cambridge University Press, Cambridge, London, p 23
17. Gupta AP, Sutherland DJ (1967) Phase contrast and histological studies of spherule cells in cockroaches. Ann Entomol Soc Am 60: 557-565
18. Harpaz I, Kislev N, Zelcer A (1969) Electron microscopic studies on hemocytes of the Egyptian cottonworm, *Spodoptera littoralis* (Boisduval) infected with a nuclear polyhedrosis virus, as compared to noninfected hemocytes I. Noninfected hemocytes. J Invertebr Pathol 14: 175-185
19. Hoffmann JA, Stoeckel ME (1968) Sur les modifications ultrastructurales des coagulocytes au cours de la coagulation de l'hémolymphe chez *Locusta migratoria*. C R Soc Biol (Paris) 162: 2257-2259
20. Hoffmann JA, Stoeckel ME, Porte A, Joly P (1968) Ultrastructure des hémocytes de *Locusta migratoria*. C R Acad Sci (D) 266: 503-505
21. Kaaya GP, Ratcliffe NA (1982) Comparative study of hemocytes and associated cells of some medically important dipterans. J Morphol 173: 351-365
22. Neuwirth M (1974) Granular hemocytes, the main phagocytic blood cells in *Calpodes ethlius* (Lepidoptera). Can J Zool 52: 783-784
23. Nittono Y (1960) Studies on the blood cells in the silkworm *Bombyx mori* (L). Bull Seri Exp Stn 16: 261-266
24. Ratcliffe NA, Gagen SJ (1977) Studies on the in vivo cellular reactions of insects: an ultrastructural analysis of nodule formation in *Galleria mellonella*. Tissue Cell 9: 73-85
25. Ratcliffe NA, Price CD (1974) Correlation of light and electron microscopic haemocyte structure in the dictyoptera. J Morphol 144: 485-498
26. Ratcliffe NA, Rowley AF (1981) Insects. In: Ratcliffe NA Rowley AF (eds) Invertebrate Blood Cells. Academic Press, London
27. Rizki MT (1957) Alterations in the haemocyte population of *Drosophila melanogaster*. J Morphol 100: 437-458
28. Rizki MT, Rizki RM (1959) Functional significance of the crystal cells in the larva of *Drosophila melanogaster*. J Biophys Biochem Cytol 5: 235-239
29. Rowley AF (1977) The role of the haemocytes of *Clitumnus extradentatus* in haemolymph coagulation. Cell Tissue Res 182: 513-524
30. Rowley AF, Ratcliffe NA (1976a) An ultrastructural study of the in vitro phagocytosis of *Escherichia coli* by the haemocytes of *Calliphora erythrocephala*. J Ultrastruct Res 55: 193-202
31. Rowley AF, Ratcliffe NA (1976b) The granular cells of *Galleria mellonella* during clotting and phagocytic reactions in vitro. Tissue Cell 8: 437-446
32. Rowley AF, Ratcliffe NA (1979) An ultrastructural and cytochemical study of interactions between latex particles and haemocytes of the wax moth *Galleria mellonella*. Cell Tissue Res 199: 127-137
33. Schmidt AR, Ratcliffe NA (1978) The encapsulation of araldite implants and recognition of foreignness in *Clitumnus extradentatus*. J Insect Physiol 24: 511-521

34. Wigglesworth VB (1973) Haemocytes and basement membrane formation in *Rhodnius*. J Insect Physiol 19: 831-844
35. Zachary D (1978) Evolution des cellules sanguines et de la musculature au cours de la métamorphose chez le Diptère *Calliphora erythrocephala*: aspects cytologiques et neuroendocrines. Thesis, L. Pasteur University, Strasbourg
36. Zachary D, Hoffmann JA (1973) The haemocytes of *Calliphora erythrocephala*. Z Zellforsch Mikrosk Anat 141: 55-73
37. Zachary D, Brehélin M, Hoffmann JA (1975) Role of the thrombocytoids in capsule formation in the dipteran *Calliphora erythrocephala*. Cell Tissue Res 162: 343-348

Chapter 5 Haemogram and Its Endocrine Control in Insects

J. P. N. Pathak[1]

CONTENTS

1 Introduction

Since the time of Kopéc [19,20] when the idea of activation hormone was produced by the neurosecretory cells of insects, the insect endocrinologists had taken much pains in exploring the various events of development and growth in insect life. The ubiquity of neurosecretion and significance of endocrine structure in metabolism of insect are now very well established. A few events of insect physiology such as moulting, tanning, excretion and maturation of gonads are well explored and understood, but the endocrine regulation of haemogram is an occult of insect physiology which needs profound investigations. Surprisingly insect endocrinologists have not developed much interest to reveal this problem. However, a great deal of data has been accumulated to explain the qualitative and quantitative changes in the haemocyte population of various species of insects during growth and development. Much has been written about the effects of injury or trauma, starvation, sex, diapause, infection or injection of particulate and nonparticulate materials on the total and differential haemocyte counts. The humoral control of insect blood cell populations was reviewed by Crossley [5]. He assumed that the changed numbers of circulating haemocytes are the result of a dynamic balance between four factors: (a) mitosis of circulating haemocytes, (b) death or fragmentation of circulating haemocytes, (c) release or retention of haemocytes at haemocytic reservoirs and (d) release of sessile haemocytes from haemocytopoietic tissues. These postulations are very convincing but we could not gather much evidence to prove the interrelationship between endocrine and the above factors.

---

1 Department of Zoology, Madhav Science College, Vikram University, Ujjain, India

Immunity in Invertebrates
Ed. by M. Brehélin

Looking to the allocation of the space in this article I want to discuss the little work which is performed in my laboratory and the other available articles which provide some convincing circumstantial assessment on the endocrine control of insect haemogram. The general studies related to the total haemocyte counts (THC) and differential haemocyte counts (DHC) were not included in this article. Readers are advised to consult the review articles of Wigglesworth [36], Jones [15,18], Arnold [1], Crossley [5], Fier [10] and Shapiro [33] for extensive information and general concept of haemogram in insects.

2 Effect of Ligature on Haemogram

The concept on the involvement of the anterior endocrine glands in the blood picture came from the work of Jones [17]. He ligatured *Rhodnius* larvae behind the head just after a blood meal and speculated that the ligature prevents hormonal circulation in the organism, therefore, the blood picture is modified. He concluded that if fifth instar *Rhodnius* is ligatured before the moulting hormones are fully secreted (i.e. 1 h after feeding to 4 days after feeding) the amount of haemolymph from a severed appendage remains abnormally high and the number of prohaemocytes and plasmatocytes is influenced. If the head is ligatured after the critical period (i.e. 8 and 14 days after feeding), then, the haemocyte picture tends to be like that of normal unligatured insects. He deduced that hormones regulate the amount of haemolymph available from an appendage and the cyclical changes in the types and number of haemocytes.

Crossley [4] observed the change in the population of phagocytic cells in *Calliphora* at the time of puparium formation and assumed that the haemocyte alterations may be under hormonal control. He further investigated this hypothesis by influencing the titre of moulting hormone by ligature and by injecting analogue crust ecdysone. Crossley [5] injected two *Calliphora* units of β-ecdysone in the feeding larvae and noted that the haemocyte population modified by a significant increase in proportion to circulating phagocytic cells. He ligatured the larvae so as to confine Weismann's ring to the anterior part of the body and found that the posterior part of the larvae could not puparate and maintain a low percentage of phagocytic cells. The percentage of phagocytic cells could be enhanced by injecting β-ecdysone (2 *Calliphora* units = 2 Cu = 0.004 mg) in the posterior part. His histogram clearly indicates the interconversion in haemocyte population due to the titre of hormone and ligature. Probably this was the first evidence which claimed the involvement of endocrine glands and hormonal level with the haemogram of insects. But the way by which β-ecdysone effects the population of phagocytes is still dubious. It may be that β-ecdysone may cause some change in the metabolic, physiologic or histologic state of the pupa which results in an increase in phagocytic cells by interconversion of haemocytes, or by release of phagocytes from haemocyte reservoirs.

Hinks and Arnold [12] have performed a few ligature experiments in *Euxoa declarata*. The last instar larvae were subjected to the four different ligatures, out of them the third ligature experiment was quite interesting and provides some proof of the role of endocrine glands in the multiplication of haemocytes. In this experiment larva was divided into three compartments with the help of two ligatures. The one i.e. anterior section has all anterior endocrine glands and both pairs of haemopoietic organs and two consecutive sections with neither head nor haemopoietic tissue. Their histogram clearly indicates that prohaemocyte and plasmatocytes are higher in the middle and posterior sections. They reported that the ligation also affected the mitotic indices in both the haemopoietic organs and the circulating haemocytes only when the head and prothorax were included in the anterior compartment, while there was minor variation between the mitotic indices of circulating haemocytes in the posterior compartment and anterior compartment in which the head and prothorax were excluded. They noted that the exclusion of the head and prothorax reduced the mitotic index by 50% in the anterior compartment. These workers further concluded that the enhanced percentage of mitosis and increase in the number of prohaemocytes and plasmatocytes in the sections where head and prothorax were included, strongly emphasizes the role of endocrine glands. They could not demonstrate which hormone is responsible for these changes but they suggested that the change in haemocyte population and mitosis index is induced by ecdysone.

3 Effect of Bleeding and Encapsulation on Haemogram

In *Carausius* [31] the mitosis index was highest between 3 to 9 h after bleeding. In *Galleria* [32] a little injury has a significant effect on the haemocyte population, THC increased within 24 h, in proportion to increases in the relative numbers of plasmatocytes, granular haemocytes, prohaemocytes and spherule cells. There was a decrease in adipohaemocytes. It is considered that injury resulting from haemorrhage whether small or large, produced similar response by elevating the mitotic index and total haemocyte population. Harvey and Williams [11] and Shapiro [32] showed that the increase in the number of haemocytes resulting from injury was largely due to an increase in plasmatocytes. Hinks and Arnold [12] considered this increase as a direct effect upon the haemopoietic organs. While Day [8] was of the opinion that the immediate increase in the number of haemocytes was due to mobilization of cells from the sedentary state and that they were replaced slowly by mitosis. A number of workers [3,7, 11,21] have concluded that wounds or increased haemorrhage usually seem to cause an increase in haemocyte number mainly due to an increase in the prohaemocyte – plasmatocyte group. In *Locusta* loss of haemocytes due to haemorrhage was restored by sudden release of undifferentiated cells from haemopoietic tissue [13]. It seems that bleeding or injury induces some factor which stimulates the mitotic activity of haemocyte and/or mobilizes the sedentary cells and/or releases the poorly differentiated cells from the haemopoietic organs. Harvey and Williams [11] suggested that such a factor is generated at the margin of the wound and is released into the blood. The distribution and quantitative changes in the blood cells of larvae of *Musca* and *Orthellia* provided some evidence that the haemocyte of these insects exhibit chemotaxis in response to parasitization [25]. Nappi [23] proposed that the cell-mediated immune reactions result in encapsulation and melanization of internal metazoan parasites brought about by changes in the normal titre of host hormone. He suggested that a hormonal imbalance affected the permeability of the cell membrane of the haemocytes to certain metabolites and brought about the premature differentiation and migration of these cells. Nappi [24] raised a question that if, indeed, the cellular immune mechanism of some insects respond to changes in the concentration of hormones, what is the origin and mode of action of the initial stimulus? He hypothesized that the stimulus originates at or very near the foreign organism or it may be a substance produced either by the parasite or by the circulating haemocytes when they come in contact with the parasite. This stimulus in turn acts directly on the brain-endocrine complex and thus affects the production of hormone (s). In response to a change in the hormone concentrations the haemocytes differentiate and migrate from areas where they are normally found in large numbers. A similar hypothesis may be proposed for the synthesis of some chemical stimulus which may be responsible for the release of cells from the haemocyte depot or from tissue surface during bleeding and the titre of hormones may act as an inducer in the formation of such a factor.

4 Effect of Ablation and Implantation of Endocrine Glands on Haemogram

The association of ring gland with the haemocytes was first suggested by Rizki [29, 30] in *Drosophila melanogaster.* He reports that specific change in cell morphology, such as the formation of pseudopodia and cell flattening only occurs at certain stages of the life cycle in *Drosophila*. He further observed that in the *ore-R* strain lamellocyte forms normally occur during the formation of puparium while in the *tumor W* mutant it advances at the time of the second moult. Experimental exclusion of that ring gland and brain resulted in an earlier appearance of lamellocyte which normalize after reintroduction of the ring gland. Rizki, however, reported further that lamellocytes may also be induced by injecting water in the larvae lacking brain and ring gland. These experiments are interpreted as arguing for hormonal control over haemocytic morphology, but on examination, it was found that the changes in haemocyte populations just before pupariation and on experimental injection of distilled water are somehow related to the process of encapsulation.

Hoffmann [14] performed certain ablation and implantation experiments with endocrine glands in *Locusta migratoria migratoriodes*. He reported that electrocoagulation of the pars intercerebralis in the male *Locusta*, increases the volume of haemolymph which was 235.5% more than the normal insects. The amount of haemocyte per mm^3 of haemolymph is less by 34.6% as compared to the normal value in electrocoagulated insects taken into consideration. However, this change in the volume of haemolymph does not modify the population of each type of haemocyte. The electrocoagulation of the pars intercerebralis in female affected the total haemocyte counts which increased abruptly with significant increase in the number of plasmatocytes. He removed the corpora allata in *Locusta* and noted a decrease in the total haemocyte count per mm^3. He also noted a strong variation in numerical value of plasmatocytes while the implantation of the corpora allata yields no significant results related to the weight of haemolymph and fresh tissue, but the total haemocytes per mm^3 increased considerably. The implantation of corpora cardiaca does not affect the weight of the insects but the total quantity of haemolymph is decreased and the concentration of haemocytes was increased. Extirpation of corpora cardiaca affected the rate of differentiation of haemocytes but the author could not come to any definite conclusion, because the effect of the ablation of corpora cardiaca on the physiology of insecta was so extensive that it was difficult to interpret.

Pathak [27] performed the extirpation and implantation of endocrine glands in *Halys dentata.* It is obvious from his extirpation experiments that in *Halys dentata* the hormonal secretion of the corpora allata influences the total haemocytes counts. The secretion of corpora allata is more effective up to the age of 6 days. Later on the count shows an abrupt and significant rise form 10, 313 ± 921 to 12, 601 ± 102. This indicates that the haemocyte population is not entirely under the influence of the corpora allata throughout the life span, and/or physiological conditions of an insect. This assumption is further strengthened by other extirpation experiments. When both the corpora cardiaca and corpora allata were removed from the insects, the count remained significantly low throughout the life span. When only the corpora cardiaca was amputated, the haemocyte count decreased significantly after the sixth day. To confirm the influence of the corpora cardiaca on the haemocyte population in *Halys dentata*, two implantation experiments were planned. In one case, the glands of 1-day-old adult insects, were transplanted into 1-day-old adults, and there was no change in the haemocyte count throughout the life of the adult. In the other case, when corpora cardiaca of 6-day-old adult insects were transplanted into 1-day-old adults, the haemocyte count increased for the first 3 days, and then became normal. Therefore, it is assumed that these glands in older adults either secrete some hormone or most probably receive a neurosecretion from the pars intercerebralis after 5 or 6 days, which in turn influences the total haemocyte count.

Implantation of corpora allata causes a significant rise in the population of haemocytes in the adults on the fourth, fifth, and sixth days. Later on these counts become normal while the implantation of brain, corpora allata and corpora cardiaca together produce a significant rise in the total haemocyte counts throughout the entire life span. The haemocyte count was also increased significantly when only a brain was implanted but the increase was not so high as when brain, corpora cardiaca and corpora allata were implanted. It was concluded that the corpora allata influenced the count during the early part of adult life, and secretion of neurosecreting cells of the brain and the corpora cardiaca influenced the THC during the later half of adult life.

The observations on the DHC of *Halys dentata* were not completed after ablation or implantation of endocrine glands. However, in the case of brain, corpora cardiaca and corpora allata implanted into adults the increase in the number of prohaemocytes and plasmatocytes was several-fold. Similarly the number of dividing cells (prohaemocytes) which was negligible in controls also increased (Pathak, unpublished data). It should also be noted here that the percentage of plasmatocytes is also higher in sham operated controls in comparison to those adults which have no injury. In view of my earlier observations on this insect it should also be taken into consideration that during all the ablation or implantation experiments the total haemocyte counts were never as high as the total haemocyte counts of heat-fixed adults [2]. These observations indicate that injury and haemorrhage influence the haemogram either by releasing the sedentary haemocytes from some depot or from tissues where they adhere. This increase if further accelerated by implanted endocrine glands. These results are very similar with the ligature experiments of Hinks and Arnold [12] where they found that ligation could influence the cell population more when the endocrine glands and haemopoietic tissues were ligated in the same compartment.

5 Interrelationship Between Ecdysis, Blood Volume and Haemogram

The cyclical changes in the population of haemocytes has been frequently demonstrated by many workers during growth and development but these studies did not provide any information on the role of hormone in the qualitative and quantitative variation of haemogram [2,14,16,37]. However, Arnold [1] pointed out that the variation in the haemogram during ecdysis may be influenced by the titre of ecdysone in the blood.

Webley [34] observed the interrelationship between THC and blood volume (BV) in *Locusta migratoria migratoriodes* and concluded that in older adults a decrease in BV is related to an increase in haemocyte population. Nittono [26] advocated the inverse relationship between the THC and BV in *Bombyx* larvae. Wheeler [35], in his studies on changes in THC in relation to blood volume during the moulting cycle of *Periplaneta americana*, calculated that the increase in THC prior to ecdysis was not significant and the decrease in THC at ecdysis itself results from a sudden but brief increase in the blood volume. He further mentioned that decrease in the absolute number of haemocytes 24 h after ecdysis is significant but does not affect the THC because the blood volume also decreases. He, therefore, concluded that as the blood volume goes up the THC falls. In *Halys dentata* [2] it has been found that there is a significant decrease in THC after ecdysis. A possible explanation for this would be that after ecdysis the blood volume falls suddenly so that a good number of haemocytes may attach themselves to the tissues and may not come into circulation. Probably, they return to the circulating blood again when the blood volume becomes normal after feeding, usually during the middle period of a stadium. In *Halys dentata* and *Bombyx mori* some recent studies were performed [28] to establish the interrelationship between blood volume, endocrine glands and total haemocyte counts.

As shown in Fig. 1 in *H. dentata*, four experiments were performed to show the effect of endocrine extracts and the solution of 5-hydroxytryptamine on the unfixed haemocyte counts. It was noted that the brain extract of 6-day-old, well-fed, hydrated adults of *H. dentata* was able to induce the excretory activity in a 1-day-old, well-fed adult, and therefore affected the blood volume and thus the total haemocyte counts increased. Similar observations were made in the corpora cardiaca extract of 6-day-old, well-fed hydrated insects, though the inducing capacity was much less than the extract of brain. The endocrine extracts of 1-day-old adults and 6-day-old dehydrated adults could not influence the process of excretion or the total haemocyte counts, probably they lack the diuretic hormone. These observations also provide a clue to the observations of Pathak [27] in which he surmised that the implantation of the corpora cardiaca of a 1-day-old adult could not influence the haemocyte counts while the implantation of glands from 6-day-old adults could enhance the haemocyte

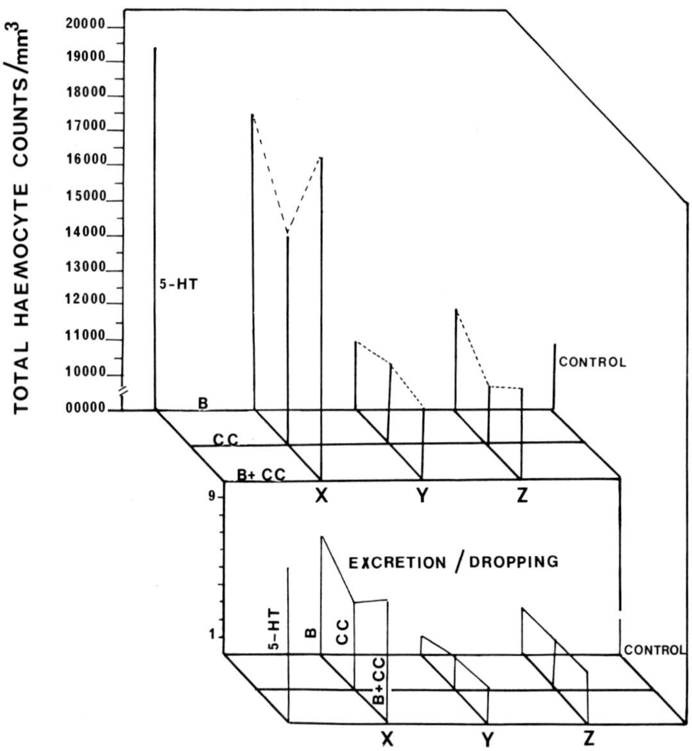

Fig. 1. Graph showing the mean value of unfixed total haemocyte counts and excretion droppings (in 2 h) in *Halys dentata* after injecting the extract of various endocrine glands and 5-hydroxytryptamine solution. Counts were made after 2 h. *X* = Extract of endocrine glands from 6-day-old hydrated adults; *Y* = Extracts of endocrine glands from 6-day-old dehydrated adults; *Z* = Extracts of endocrine glands from 1-day-old adults; *B* = Brain; *CC* = corpora cardiaca

counts for the first 3 days. It seems that in 6-day-old, well-fed insects, neurosecretory cells of brain secrete diuretic hormone which reduces the blood volume and thus an increase in the total haemocyte counts was noted. Probably, this increase has not affected the absolute number of cells as the total haemocyte counts never crossed the number in heat-fixed specimens [2].

In the present study it was noted that the extract of corpora cardiaca (6-day-old adult insect) was able to influence the excretion rate and also the haemocyte counts. Therefore, it seems that the secretion of corpora cardiaca has some diuretic factor which stimulates the fluid movement through the Malpighian tubules and inhibits its reabsorption by the rectum [32] and therefore more water is excreted to reduce the haemolymph volume. However, such a diuretic factor is not released by the corpora cardiaca of a 1-day-old adult, *Halys dentata*. It may be that for the secretion of this factor some intrinsic stimuli may develop after the growth of 6 days.

In the spinning larvae of *B. mori* similar studies were performed to show the interrelationship between the blood volume and the unfixed haemocyte counts. It was noted that due to the influence of the diuretic hormone secreted from either neuro-

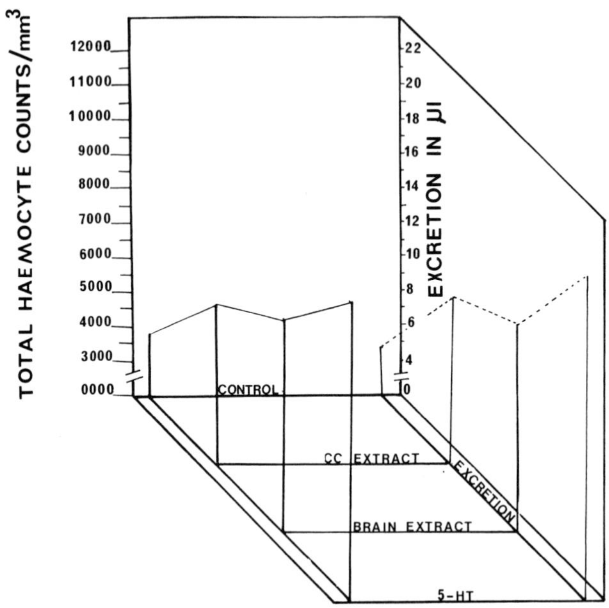

Fig. 2. Graph showing the mean values of unfixed total haemocyte counts and fluid excretion in μl (in 2 h) in the spinning larvae of *Bombyx mori* after injecting the extract of various endocrine glands and 5-hydroxytryptamine solution. Counts were made after 2 h

secretory cells of brain or from corpora cardiaca the fluid from the haemolymph is excreted which affects blood volume and thus the total haemocyte counts were increased in comparison to control (Fig. 2). A similar relationship has also been proven by the diuretic drug i.e. 5-hydroxytryptamine.

Essawy et al. [9] studied the total and differential haemocyte counts and combined them with blood volume in order to estimate the changes which occur in the haemocyte picture in the last larval stage of *Heliothis armigera*. They noted that with the increase of blood volume at 96 h the absolute number of haemocytes increased but at this time the THC is decreased. They also studied the DHC and noted that the number of cells per cubic mm haemolymph shows that the peak of prohaemocytes is at 24 h. The peak of spherule cells, coagulocytes and oenocytoids are at 48 h, while the peak of plasmatocytes is at 72 h. The peak of DHC of total plasmatocytes which occurs at 72 h corresponds to the end of the feeding period, to the maximum of the haemolymph osmolarity and the second peak of ecdysone. If the head of the larva is ligated before this peak of plasmatocytes the prothoracic gland is found ultrastructurally active, but the periphery of the cells exhibit numerous digitations which establish desmosome-like junctions. In this state the periphery of the gland seems to be impermeable for the exchanges between the gland and haemocytes. They also found the peak of coagulocytes at 120 h i.e. 24 h before the peak of maximum ecdysone. Therefore they proposed the hypothesis that the plasmatocyte contains material perhaps coming from the brain, which modifies the permeability of the gland.They further noted that the DHC of plasmatocytes and the coagulocytes are also influenced by the titre of ecdysone.

6 Conclusions

1. The ligature experiments provide evidence that the anterior endocrine glands affect the haemogram in insect. These endocrine glands influence the total and differential haemocyte population in several ways viz. (a) interconversion of haemocyte i.e. formation of phagocytes and/or lamellocytes; (b) increasing the mitotic index; (c) releasing the haemocytes from haemopoietic tissue; and (d) mobilizing the sessile and adhering cells. These studies also provide circumstantial evidence for the involvement of ecdysone in the above activities.

2. The studies related to injury, bleeding and encapsulation provide evidence of the involvement of endocrine glands in increasing the mitotic index, release of poorly differentiated cells from haemopoietic tissue and release of cells from adhering tissue surface. But in these reactions some mediator is involved. It is suggested that wounds, haemorrhage or parasites produce some unknown factor which affects the neuroendocrine system of insect in producing more hormones and influencing the haemocytes.

3. The ablation and implantation studies of endocrine structures indicate that endocrine glands influence the haemogram positively, but how a particular endocrine gland impels the population of haemocytes needs further investigations.

4. There is a definite interrelationship between blood volume and haemocyte population. The blood volume is under the control of several factors and one of them is hormonal. The diuretic hormone or its analogue, 5-hydroxytryptamine, may influence the haemolymph volume of an insect suddenly resulting in an increase of total haemocyte counts.

7 Prospects for Future Work

The interrelation between endocrine glands and haemogram and/or immune system in insects is a fascinating topic of neuroendocrinology in insects. I am afraid to say that the studies which are performed to elucidate this interaction are insufficient. As the endocrine glands are capable of influencing all the physiological activities of insects, studies should be performed with one particular gland to observe its effect on the various metabolical and physiological activities of the insect in relation to haemocytes to show the effect in the multiplication and differentiation of various circulating cells.

The existence of unique processes in insects which has led to the discovery of many hormonal reactions also suggests the undertaking of certain studies with isolated and artificially synthesized hormones. The burgeoning literature on insect neuroendocrinology is now able to provide information on hormonal analogues. We should use them exogenously and endogenously to observe their effects on the haemocyte population. The recent techniques developed in the field of isolation of haemocytes in pure form (see Chap. 1) will help the future investigators to observe the impact of endocrine secretion on the haemocytes in vitro. I feel such studies will also provide a clue to the feedback mechanism if it exists between the endocrine glands and haemocytes.

References

1. Arnold JW (1974) The haemocytes of insects. In: Rockstein M (ed) The Physiology of Insecta, 2nd edn. Academic Press, New York
2. Bahadur J, Pathak JPN (1971) Changes in the total haemocyte counts of the bug, *Halys dentata* under certain conditions. J Insect Physiol 17: 329-34
3. Clark RM, Harvey WR (1965) Cellular membrane formation by plasmatocytes of diapausing *cecropia* pupae. J Insect Physiol 11: 161-76
4. Crossley AC (1965) Transformations in the abdominal muscles of the blue blow-fly *Calliphora erythrocephala* (Meig) during metamorphosis. J Embryol Exp Morphol 14: 375-98
5. Crossley AC (1968) The fine-structure and mechanism of breakdown of larval intersegmental muscles int the blowfly *Calliphora erythrocephala*. J Insect Physiol 14: 1389-1407
6. Crossley AC (1975) The cytophysiology of insect blood. Adv Insect Physiol 11: 117-222
7. Davis RP, Schneiderman HA (1960) An autoradiographic study of wound healing in diapausing silkworm pupae. Anat Rec 137-348
8. Day MF (1952) Wound healing in the gut of the cockroach *Peripleneta*. Aust J Sci Res (B) 52: 282-289
9. Essawy M, Maleville A, Brehélin M (1984) Evolution of haemogram during the larval development (last instar) of *Eliothis armigera*. Invertebr Immunol Conf 17-29 Sept 1984, Montpellier, France. Dev Comp Immunol (in press)
10. Fier D (1979) Multiplication of haemocytes. In: Gupta AP (ed) Insect haemocytes, 1st edn. Cambridge University Press, Cambridge
11. Harvey WR, Williams CM (1961) The injury metabolism of the cecropia silkworm I. Biological amplification of the effects of localized injury. J Insect Physiol 7: 81-99
12. Hinks CF, Arnold JW (1977) Haemopoiesis in Lapidoptera II. The role of the haemopoietic organs. Can J Zool 55: 1740-1755
13. Hoffmann JA (1969) Etude de la récupération haémocytaire après hémorragies expérimentales chez l'orthoptère *Locusta migratoria*. J Insect Physiol 15: 1375-1384
14. Hoffmann JA (1970) Régulations endocrines de la production et de la différenciation des hémocytes chez un insecte l'orthoptère: *Locusta migratoria migratoroides*. Gen Comp Endocrinol 15: 198-219
15. Jones JC (1962) Current concepts concerning insect haemocytes. Am Zool 2: 209-246
16. Jones JC (1965) The haemocytes of *Rhodnius prolixus* Stal. Biol Bull (Woods Hole) 129: 282-294
17. Jones JC (1967) Effect of repeated haemolymph withdrawals and of ligaturing the head on differential counts of *Rhodnius prolixus* Stal. J Insect Physiol 13: 1351-1360
18. Jones JC (1970) Hemocytopoiesis in insects. In: Gordon AS (ed) Regulation of hemopoiesis. Appleton, New York
19. Kopéc S (1917) Experiments on metamorphosis of insects. Bull Int Acad Sci Cracovie (B): 57-60
20. Kopéc S (1922) Studies on the necessity of the brain for the inception of insect metamorphosis. Biol Bull (Woods Hole) 42: 323-342
21. Lea MS, Gilbert LI (1966) The hemocytes of *Hyalophora cecropia* (Lepidoptera) J Morphol 118: 197-216
22. Mordue W (1969) The hormonal control of Malpighian tubue and rectal function in the desert locust *Schistocerca gregaria*. J Insect Physiol 15; 273-285
23. Nappi AJ (1973a) The role of melanization in the immune reaction of larvae of *Drosophila algonquin* against *Pseudeucoila bochei*. Parasitology 66: 23-32
24. Nappi AJ (1973b) Hemocytic changes associated with incapsulation and melanization of some parasites. Exp Parasitol 33: 285-302
25. Nappi AJ, Stoffolano JC Jr (1971) *Heterotylenchus autumnalis*: Hemocytic reactions and capsule formation in the host, *Musca domestica*. Exp Parasitol 29: 116-125
26. Nittono Y (1960) Studies on the blood cells in the silkworm, *Bombyx mori* (L). Bull Seric Exp Stn 16: 171-266

27. Pathak JPN (1983) Effect of endocrine glands on the unfixed total haemocyte counts of the bug, *Halys dentata*. J Insect Physiol 29: 91-94
28. Pathak JPN (1984) Effect of endocrine extracts and 5-HT on the unfixed total haemocyte counts of *Halys dentata* and spinning larvae of *Bombyx mori*. Invertebr Immunol Conf 17-20 Sept 1984, Montpellier, France. Dev Comp Immunol (in press)
29. Rizki MT (1957a) Alterations in the hemocyte population of *Drosophila melanogaster*. J Morphol 100: 437-458
30. Rizki MT (1957b) Tumor formation in relation to metamorphosis in *Drosophila melanogaster*. J Morphol 100: 459-472
31. Rooseboom A (1937) Contribution à l'étude de la cytologie du sang de certains insects, avec quelques Considérations générales. Arch Neerl Zool 2: 432-559
32. Shapiro M (1968) Changes in the hemocytes population of the wax moth *Galleria mellonella* during wound healing. J Insect Physiol 14: 1725-1733
33. Shapiro M (1979) Changes in hemocyte population. In: Gupta AP (ed) Insect Hemocyte, 1st edn. Cambridge University Press, Cambridge, pp 475-523
34. Webley DP (1951) Blood cell counts in the African migratory locust (*Locusta migratoria* or *migratorioides* Reiche and Fairmaire) Proc R Entomol Soc Lond Ser Gen Entomol A 26: 25-37
35. Wheeler RE (1963) Studies on the total hemocyte count and hemolymph volume in *Periplaneta americana* (L) with special reference to the last moulting cycle. J Insect Physiol 9: 223-235
36. Wigglesworth VB (1959) Insect blood cells. Annu Rev Entomol 4: 1-16
37. Yeager JF (1945) The blood picture of the southern armyworm (*Prodenia eridania*). J Agric Res China 71: 1-40

Section 2 The Molecules of Immunity and the Immunorecognition Process

Chapter 6 Antibacterial Immune Proteins in Insects – A Review of Some Current Perspectives

H. G. Boman[1], I. Faye[1], P. v. Hofsten[1], K. Kockum[1]
J.-Y. Lee[1], K. G. Xanthopoulos[1], H. Bennich[2]
Å. Engström[2], B. R. Merrifield[3], and D. Andreu[3]

CONTENTS

1 Introduction

More than 10^6 insect species are recognized in the literature and estimates indicate that the number of individual insects is as high as 10^{18} [32]. These high numbers predict that insects will have very effective immune systems, if not they would not be as numerous as they are. The large number of insect species also creates a need for the selection of model systems. Many investigators have used larvae of the greater wax moth, *Galleria mellonella* and vaccination of this insect was reported already 60 years ago. However, hemolymph from rapidly growing larvae contains a large number of proteins and when comparing *Galleria* to diapausing pupae of *Hyalophora cecropia* we found the latter superior both for immunological work and for purification of proteins [11]. The cecropia pupae are large, they weigh 5–10 g, and they contain 1–2 ml of hemolymph. When a diapausing pupa is immunized it turns on predominantly the genes for immunity while the rest of the animal remains in a dormant state. Immunized pupae of cecropia are, therefore, a system for biological enrichment of the RNA and the proteins which are synthesized from the genes for immunity. We have taken advantage of this fact both in the purification of 15 different immune proteins and in the isolation of immune RNA, later to be used for the preparation of a cDNA bank. During the last 5 years we have carried out parallel sequence work on the protein and DNA level. This program has so far produced the complete amino acid sequences for five cecropins, one lysozyme, and one attacin. In addition, we have obtained cDNA

1 Department of Microbiology, University of Stockholm, S-106 91 Stockholm, Sweden
2 Department of Immunology, BMC, University of Uppsala, S-751 23 Uppsala, Sweden
3 Rockefeller University, New York, New York 10021, USA

Immunity in Invertebrates
Ed. by M. Brehélin

sequences corresponding to one lysozyme, one cecropin, and the two major forms of attacin. In cecropins, work on both protein and cDNA level yielded structural information that was used in the synthetic work. A summary of these results is included in this brief and selective review and in another recent symposium paper [5]. A more complete review of cellular and humoral immunity in insects was provided by Götz and Boman [9].

2 Early Work on Insect Immunity

The early work on insect immunity was limited to vaccination studies and it was taken for granted that insects produced "antibodies". Later on when mammalian antibodies were better understood the search in insects led to the discovery of insect lectins [3]. The first in vitro demonstrations of antibacterial activities in hemolymph from immunized insects were the detailed studies of Briggs [6] and Stephens [30]. The first antibacterial factor to be identified in insect hemolymph was lysozyme and it has been claimed that this enzyme is the main antibacterial factor responsible for immunity of vaccinated insects [22]. We now know that this is not correct because insects can eliminate many lysozyme-resistant bacteria [4]. The lysozymes from *Galleria mellonella* and *Bombyx mori* were purified by Powning and Davidson [24] and the *Galleria* enzyme has continued to interest other investigators [16,17].

3 The Primary Structure of the Cecropia Lysozyme

We isolated the cecropia lysozyme in connection with the purification of cecropin A and B [13]. This enzyme was used by Engström et al. (EMBO J. 4: 2119-2122, 1985) for the elucidation of the complete amino acid sequence. Our DNA work was initiated when most of the protein structure was already known. It was, therefore, possible to construct a synthetic oligonucleotide probe and use this for the isolation of a clone, pCP701, containing an insert with about 400 base pairs. This clone was sequenced and found to contain the full structural information for the mature cecropia lysozyme (Fig. 1). A comparison with the protein sequence showed two possible discrepancies, namely, at positions 15 and 66. These were resolved as follows.

In the beginning of our purification work on the cecropia lysozyme we obtained only one peak. However, as time went on (and other dealers were used for our supply of pupae) we usually obtained two peaks. Sequence studies on these two lysozymes revealed one component with Arg-15 and one with Leu-15. Furthermore, in the sequence work, lysozyme with Leu-15 always gave two signals at position 66, one for serine and one for threonine. It is, therefore, likely that there are three allelic variants in the cecropia population with the following alternatives (I) Arg-15, Ser-66; (II) Leu-15, Ser-66, and (III) Leu-15, Thr-66. These three variants can all be derived from each other by point mutations because argenine-leucine and serine-threonine replacements can be obtained by transversion of a single base.

```
                                        25                                              50
TGC CGT TCG TGG CAG TTC GCT TTG CAT TGC GAT GCG AAA CGT TTC ACG AGA TGC GGG TTA
Cys Arg Ser Trp Gln Phe Ala Leu His Cys Asp Ala Lys Arg Phe Thr Arg Cys Gly Leu
       -10                                             -1 || 1
------------------- Leader peptide -----------------------||------------->  Lysozyme

                    75                                              100
GTG CAG GAG CTT AGG AGA CGA GGC TTC GAT GAA ACT TTG ATG AGT AAC TGG GTC TGC CTT
Val Gln Glu Leu Arg Arg Arg Gly Phe Asp Glu Thr Leu Met Ser Asn Trp Val Cys Leu
     10                                      20

 125                                     150                                     175
GTC GAG AAC GAA AGC GGA CGG TTT ACC GAT AAA ATC GGT AAA GTT AAC AAG AAC GGA TCT
Val Glu Asn Glu Ser Gly Arg Phe Thr Asp Lys Ile Gly Lys Val Asn Lys Asn Gly Ser
     30         ---                          40

                                         ┌──────── Probe ────────┐
                         200             │                       │ 225
CGA GAC TAC GGC CTC TTC CAG ATC AAT GAC AAA TAC TGG TGC AGT AAG GGA TCC ACT CCT
Arg Asp Tyr Gly Leu Phe Gln Ile Asn Asp Lys Tyr Trp Cys Ser Lys Gly Ser Thr Pro
    ---
     50                                      60

        250                                 275                                 300
GGA AAG GAT TGC AAC GTG ACT TGT AAT CAG CTA CTG ACT GAC GAC ATT AGC GTG GCA GCT
Gly Lys Asp Cys Asn Val Thr Cys Asn Gln Leu Leu Thr Asp Asp Ile Ser Val Ala Ala
     70                                      80

                                        325                                 350
ACG TGC GCG AAG AAG ATT TAC AAA CGC CAC AAG TTT GAC GCT TGG TAC GGA TGG AAA AAT
Thr Cys Ala Lys Lys Ile Tyr Lys Arg His Lys Phe Asp Ala Trp Tyr Gly Trp Lys Asn
     90                                     100

                    375                                 400
CAC TGT CAA CAT GGA CTG CCA GAT ATT AGC GAC TGT TAG  AGACGACTTATTATAGCCTTC
His Cys Gln His Gly Leu Pro Asp Ile Ser Asp Cys Stop
    110                                     120
```

Fig. 1. Nucleotide and deduced amino acid sequences of the cecropia lysozyme. Part of the leader peptide is indicated by a *line under the respective sequence*. The principal structure of the oligonucleotide probe (without the alternative wobble bases) is indicated by a *line above the respective sequence*. The active site of the enzyme, residues Glu-32 and Asp-50, are *underlined*

4 Cecropins, Analogs, and Precursors

Cecropins were discovered in 1979 when we succeeded in their separation from the cecropia lysozymes. Once separated, cecropin A and B were isolated simultaneously with the lysozyme [13]. Two years later we found cecropin D as well as three minor forms believed to be precursors [14]. In collaboration with the Shanghai Institute of Biochemistry, we have also isolated cecropin D and B from the Chinese oak silk moth, *Antheraea pernyi* [26]. Figure 2 shows the structure of these five cecropins with the three sequences from *Hyalophora cecropia* (H.c.) in the middle flanked by the corresponding two forms of the Chinese oak silk moth, *Antheraea pernyi* (A.p.). It is clear that all cecropins are, in principle, similar with a strongly basic N-terminal region and a long hydrophobic stretch in the C-terminal half. The high degree of homology shown by the five cecropins (boxed-in residues in Fig. 2) also suggests that they have evolved through gene duplications.

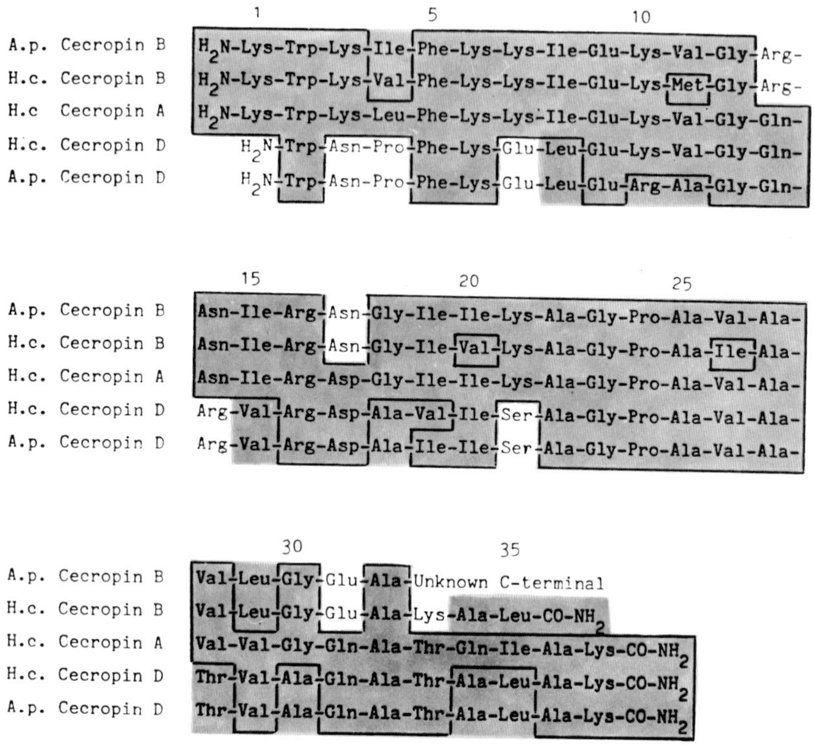

Fig. 2. Sequences for cecropin A, B, and D from *Hyalophorae cecropia (H.c)* and cecropin B and D from *Antheraea pernyi (A.p)*. Identical amino acid residues are *boxed in* and essentially similar residues are *shaded*

In the N-terminal region polar and hydrophobic side chains are interspaced in a regular pattern. Steiner [28] and Merrifield et al. [21] observed that this amino acid distribution is likely to produce an amphipathic alpha helix. Such structures have often been implicated in membrane activities and the cecropins are indeed strongly lytic against a variety of bacteria [29]. Cecropins will also lyse artificial liposomes (Steiner, unpublished results).

To confirm the sequences and to elaborate on the relationship between structure and function we initiated a program for solid phase synthesis of cecropins. A beginning was made with the synthesis of cecropin A (1-33) at that time thought to be the full cecropin A [21]. This was followed by the synthesis of the complete cecropin A and some truncated analogs [1]. It was also shown that cecropin A has a C-terminus blocked by an amide group. As a continuation a series of analogs of cecropin A has been synthesized in which residues 2, 6, and 8 have been altered in such a way that the nature of the side chain has been changed from hydrophobic to hydrophilic or vice versa [2]. We have also replaced residues 4 or 8 by Pro in order to break the alpha helix.

Table 1. Lethal concentrations (μM) of cecropin A and synthetic analogs[a]

Cecropin A analog	*E. coli* D21	*P. aeruginosa* OT97	*B. megaterium* Bm11	*M. luteus* Ml11
(1-37) Nat.	0.35	2.6	0.60	1.4
(1-37) Syn.	0.28	1.5	0.56	1.5
(2-37)	0.37	8.6	1.5	7.3
(3-37)	2.6	90	13	> 110
(1-33)	0.43	13	24	27
(Phe-2)	0.34	3.5	0.78	7.4
(Glu-2)	3.2	170	39	> 170
(Pro-4)	0.36	8.1	11	87
(Leu-6)	0.56	120	0.78	7.3
(Glu-6)	0.58	34	2.2	4.7
(Pro-8)	0.50	15	31	80

[a] Thin agar plates were prepared seeded with the respective test bacteria. Small wells were punched in the plates and loaded with a dilution serie of each sample. After overnight incubation at 30°C the inhibition zones were recorded and lethal concentrations calculated as described [15]

The results in Table 1 show that two of the truncated analogs (2-37) and (1-33), are fully active against *E. coli* while the complete molecule is needed for a good activity against all of the test bacteria. The Phe-2 analog showed only a five fold reduction in activity against *M. luteus*, while the Glu-2 analog showed a ten fold reduction of the activity against *E. coli* and almost no activity against the three other test bacteria. Also the truncated analog (3-37) showed a reduced activity. We can, therefore, conclude that Trp-2 is an essential residue. It is also evident that the two analogs with Pro-4 and Pro-8 are fully active against *E. coli*. Thus, there cannot be an obligate need for an extended alpha helix. However, in these cases there was no activity against *M. luteus* so the broad spectrum was lost when the alpha helix was disrupted.

It turned out to be difficult to find lysozyme and cecropin clones using messenger selection and immunoprecipitation. With a synthetic probe three cecropin clones were found of which the two larger were cecropin B clones [12]. The sequence data in Fig. 3 brought us three new pieces of information. Firstly, clone pCP902 contains a precursor sequence of 26 amino acid residues in the N-terminus which is not present in the mature cecropin B. The signal peptide probably makes up the first 22 of these 26 amino acids, leaving a Pro-containing tetrapeptide before the start of the mature cecropin B. This tetrapeptide is identical to the first four amino acid residues of the prosequence of melittin [19].

Secondly, the amide group in the C-terminus of the mature cecropin B must be derived from the Gly residue which terminates the coding part of the DNA sequence. The mechanism of amidation is, thus, analogous to the one found for melittin and several brain hormones. We can, therefore, conclude that in order to obtain the mature cecropin B, the precursor molecule will have to be processed in two to three steps at both ends.

Thirdly, the mature cecropin B sequence encoded by the cDNA clones is lacking two amino acid residues present in the tentative structure previously published [29]. That the cDNA structure is correct was shown by chemical synthesis of cecropin B

```
                         25                                50
TGTTTACCTATATATCTAAATTTAATATTTCGTTTATTAAAATTTAATATATTTCGACG ATG AAT TTC
                                                            Met Asn Phe
                                                            |___-25____

      75                          100
TCA AGG ATA TTT TTC TTC GTG TTC GCT TTG GTT CTG GCT TTG TCA ACA GTT TCG
Ser Arg Ile Phe Phe Phe Val Phe Ala Leu Val Leu Ala Leu Ser Thr Val Ser
__________-20____________ Leader peptide _________-10__________________

                                                 ┌── Probe ──┐
125                           150                              175
GCT GCA CCG GAG CCG AAA TGG AAA GTC TTC AAG AAA ATT GAA AAA ATG GGT CGC
Ala Ala Pro Glu Pro Lys Trp Lys Val Phe Lys Lys Ile Glu Lys Met Gly Arg
-5_______________-1||_1____________5_______________10__________________

                                200                               225
AAC ATT CGA AAC GGT ATT GTC AAG GCT GGA CCA GCG ATC GCG GTT TTA GGC GAA
Asn Ile Arg Asn Gly Ile Val Lys Ala Gly Pro Ala Ile Ala Val Leu Gly Glu
_____15___ Cecropin B ___20_______________25____________________30_____

                          250
GCC AAA GCG CTA GGA TAA  ATTAATTTTAATTT
Ala Lys Ala Leu Gly Stop
_________________35 NH2
```

Fig. 3. Nucleotide sequence of the inserts in clones pCP901 and pCP902. The structure of the leader peptide and the mature cecropin B molecule are indicated by *lines under the respective parts* of the amino acid sequence. The principal structure of the probe (without alternative wobble bases) is indicated by *lines above the base sequence*

with the structure given in Figs. 2 and 3 as well as by further structural studies on the C-terminal tetrapeptide [12].

5 The Two Main Forms of Attacin

The attacins were first isolated by molecular sieving as an antibacterial fraction with molecular weight considerably larger than the cecropins [15]. Subsequent studies revealed as many as six different components (A–F) which could be fractionated according to the isoelectric point. To our surprise they all turned out to react with antisera prepared against our immune protein P5 isolated several years earlier [25]. At that time we had not been able to find any antibacterial activity of P5, a discrepancy which later could be traced down to the type of assay used. An Ouchterlony plate revealed that all six attacins shared one antigenic determinant, while another one was common only to attacins E and F, the two acidic forms. The N-terminal sequences for five of the attacins indicated that the three basic forms all have similar sequences, while the two acidic forms are identical, but slightly different from the basic [15]. These data strongly suggested the existence of only two different genes, one for the basic and one for the neutral or acidic form. This was borne out by the isolation and sequencing of clones pCP517 and pCP521 [18]. Figure 4 shows that the two main attacins are very similar with as much as 79% homology on the amino acid level. On the DNA level the homology is 76% for the coding region, in contrast to only 36% in the region beyond the stop signal. Thus, as in the case of the cecropins, it seems likely that the attacins have arisen through gene duplications.

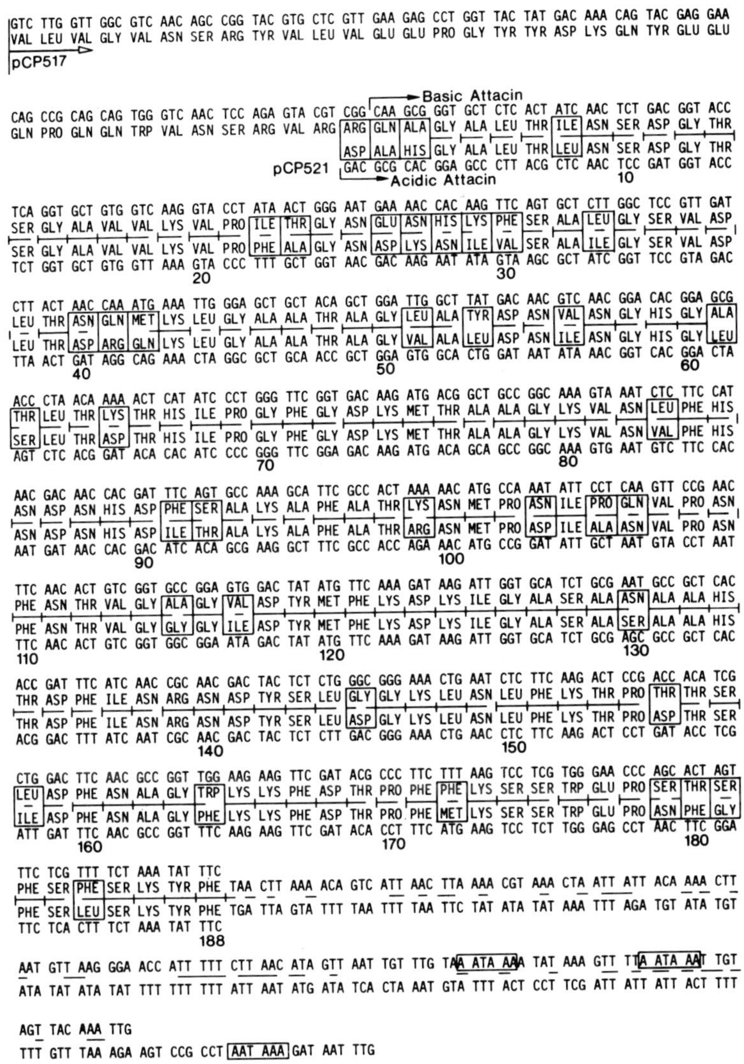

Fig. 4. Nucleotide sequences of the inserts in two attacin clones pCP517 (*uppermost line*) and pCP521 (*bottom line*). The amino acid sequence belonging to pCP517 is given below the nucleotide sequence, for pCP521 above. The line between the amino acid sequences is *broken* when the nucleotide sequences differ. Amino acid substitutions and the signals for polyadenylation are *boxed* in. The *numbers* indicate amino acid positions for both attacins

Engström et al. [7] worked out the full amino acid sequence for attacin F. There is a complete agreement between these protein data and amino acid residues 1-184 deduced from pCP521. However, pCP521 codes for an extended protein of 188 amino acid residues (attacin E). The difference corresponds to a tetrapeptide, Ser-Lys-Tyr-Phe, which is also coded for in the clone for the basic attacin. Since this peptide contains one positive charge, an incomplete proteolytic removal could account for four attacins with different charges. It cannot be decided at present if such a processing is the result of an artificial proteolytic cleavage or if it has a natural function.

In the case of pCP517 we obtained 36 amino acid residues of the leader sequence. Since neither the length nor the composition of this region is typical for a signal sequence, we believe attacins to be made from a pre-pro-form. Similar conclusions had earlier been reached from the size of the proteins produced by in vitro translation of immune mRNA [20]. In addition, there may be a further trimming also at the N-terminus of the basic attacin because evidence for a pyroglutamate group was found at the N-terminus of this molecule [7]. We, therefore, believe that the mature protein starts with the Gln residue indicated in Fig. 4. A partial cyclization of this residue would involve a partial loss of one positive charge. Thus, together with the removal of the tetrapeptide at the C-terminus it is possible by different steps of processing to account for all six forms of the attacins observed.

6 How Common are Cecropins and Attacins?

The cecropins and the attacins were first defined by their separations, especially the acidic electrophoresis in combination with an antibacterial assay on top of the gel [13]. With this technique cecropin-like substances were demonstrated in seven other Lepidoptera [11]. However, it was emphasized that ultimately it will be necessary to establish the identity of a cecropin by sequence analysis. The only insect for which this so far has been done is the chinese oak silk moth in which the D-form was the dominating cecropin [26]. However, at a recent meeting Dunn reported an incomplete sequence for cecropin D from *Manduca sexta* (Symposium arranged by the American Society of Zoologists, Denver, Dec. 29, 1984). Judging from separation properties and amino acid analysis it seems likely that cecropins are present in the meat-fly *Sarcophaga peregrina* [23]. Japanese and Chinese investigators have also reported antibacterial substances in *Bombyx mori* which very likely are cecropins (Kikuchi at a Naito Foundation Symposium and Wong and Qu in Chinese). Other preliminary data indicate antibacterial compounds also in *Drosophila* (Flyg and Dalhammar, personal communication; Robertson and Postlethwait, this symposium) as well as in *Locusta* and *Calliphora* (Hoffmann et al., this symposium).

We have been asked if in retrospective it would be possible to identify any of the antibacterial substances described in the pioneering work of Stephens and Marshall [31], Gingrich [8], and Hink and Briggs [10]. Unfortunately this is difficult because they all arrived at some misleading or erroneous conclusions, probably because they performed qualitative tests on material that was too impure. However, since both Stephens and Marshall [31] and Hink and Briggs [10] used *Galleria* larvae and since

Galleria contains several cecropin-like substances [11], it seems quite likely they were present also in the earlier work.

7 Discussion

The cecropia moth has three different cecropins and at least two different attacins. So far we have not been able to document any real differences in function between these multiple antibacterial factors. In general, cecropin B is slightly more potent than the A-form, while cecropin D has a more narrow antibacterial spectrum. However, in no case do we have a bacterium on which only a single factor acts. This raises the question of the survival value of multiple forms of very similar molecules. One alternative is that they simply represent proteins " in the middle" of an evolution toward separate functions. A second possibility is of course that both cecropins and attacins have separate target organisms which we just have not found. A third alternative, which perhaps is the most likely one, is that each of the cecropins and attacins have separate targets on most of the organisms on which they act. If so, this would in itself provide a survival value to the insect because it would make it virtually impossible for a susceptible bacterium to produce mutants which are resistant to the humoral immunity of an insect. One line of evidence supporting this alternative was obtained when we isolated mutants of *E. coli* which showed an increased susceptibility to cecropin D, but an unaltered response to the A and B forms [27]. In addition, it is clear that during the evolution of cecropia and the Chinese oak silk moth, the only mutations which have been tolerated in cecropin D are strictly conservative replacements (see Fig. 2). Also this fact speaks for a high survival value of the intact structure of cecropin D.

The exoskeleton of insects is an effective barrier against invasions of different microbes. Still, infections occur when insects are wounded and when parasite attack. However, the most common route of infection is through the gut. A special type of infection may occur during metamorphosis when the gut is dissolved by histolytic enzymes and the intestinal flora is released into the hemolymph. This would require an immune system with the capacity to eliminate a sizeable number of bacteria within a very short time and disregarding the type of bacteria that is released. The only demand for specificity would be to avoid self-destruction and this seems to be fulfilled by the combined effects of the cecropins and attacins. The role of the lysozyme would be to dissolve the murein sacculus which is left over as an empty bag when the membranes have been ruined by the detergent-like action of the cecropins.

The insects have "avoided" the complicated method of self-recognition which is used by the vertebrates. The ideal immune substances for an insect would be the ones which have the widest action against foreign intruders without causing any self-destruction. It is, therefore, interesting to compare immune substances like cecropins with a toxin such as melittin which in an ideal situation should meet rather similar demands. Both cecropins and melittin are short polypeptides containing a strongly basic part and a long stretch of hydrophobic amino acid residues. However, the polarity is reversed because in the cecropins the N-terminal region is basic, while in melittin the C-terminal is positively charged. Both kinds of molecules have also alpha-amidated

C-terminal ends formed from glycine residues. Both proteins are also made as precursor molecules, presumably in order to minimize self-destruction that could be caused by very high concentrations at the site of synthesis. Still, the mature molecules differ in their specificity: both will lyse bacteria, but only melittin will lyse eukaryotic cells [29]. Moreover, the antibacterial spectrum of melittin is narrow [4] so it could hardly qualify as an immune substance even if it were not toxic to insect cells. It is, therefore, possible that a synthetic program exploring the properties of cecropin-melittin hybrids would yield information concerning the specificity of these molecules.

Insect immunity is an interesting problem in its own rights. Besides it is also a very useful model system for the study of selective gene activation in an eukaryotic organism. In order to pursue both these problems we initiated our gene cloning program. The first phase – the exploration of a cDNA bank – is now finished and the results are summarized at this meeting.

Acknowledgements. This work was supported in Stockholm by grants from the Swedish Natural Science Research Council, in Uppsala by grants from the Swedish Medical Research Council and in New York by the U.S. Public Health Service. K.G.X. was a holder of a fellowship from "Alexander S. Onassis" Public Benefit Foundation and D.A. was a fellow of the Fundación Juan March, Madrid.

References

1. Andreu D, Marrifield RB, Steiner H, Boman HG (1983) Solidphase synthesis of cecropin A and related peptides. Proc Natl Acad Sci USA 80: 6475-6479
2. Andreu D, Merrifield RB, Steiner H, Boman HG (1985) N-terminal analogs of cecropin A: Synthesis, antibacterial activity and conformational properties. Biochemistry 24: 1683-1688
3. Bernheimer AW (1952) Hemagglutinins in caterpillar bloods. Science 115: 150-151
4. Boman HG (1982) Humoral immunity in insects and the counter defence of some pathogens. Zentralbl Bakteriol Suppl 12: 211-222
5. Boman HG, Faye I, v Hofsten P, Kockum K, Lee J-Y, Xanthopoulos KG, Bennich H, Engström Å, Merrifield RB, Andreu D (1985) On the primary structure of lysozyme, cecropins and attacins from *Hyalophora cecropia*. Dev Comp Immunol 9: 551-558
6. Briggs JD (1958) Humoral immunity in lepidopterous larvae. J Exp Zool 138: 155-188
7. Engström Å, Engström P, Tao Z-j, Carlsson A, Bennich H (1984) Insect immunity. The primary structure of the antibacterial protein attacin F and its relation to two native attacins from *Hyalophora cecropia*. EMBO J 3: 2065-2070
8. Gingrich RE (1964) Acquired humoral immune response of the large milkweed bug, *Oncopeltus fasciatus* (Dallas), to injected materials. J Insect Physiol 10: 179-194
9. Götz P, Boman HG (1985) Insect immunity. In: Kerkut GA, Gilbert LI (eds) Comprehensive insect physiology, biochemistry and pharmacology, Vol 3. Pergamon Oxford, pp 453-485
10. Hink WF, Briggs JD (1968) Bactericidal factors in haemolymph from normal and immune wax moth larvae, *Galleria mellonella*. J Insect Physiol 14: 1025-1034
11. Hoffmann D, Hultmark D, Boman HG (1981) Insect immunity: *Galleria mellonella* and other Lepidoptera have Cecropia-P9-like factors active against Gram negative bacteria. Insect Biochem 11: 537-548
12. Hofsten v. P, Faye I, Kockum K, Lee J-Y, Xanthopoulos KG, Boman IA, Boman HG, Engström Å, Andreu D, Merrifield RB (1985) Molecular cloning, cDNA sequencing and chemical synthesis of cecropin B from *Hyalophora cecropia*. Proc Natl Acad Sci USA 82: 2240-2243

13. Hultmark D, Steiner H, Rasmuson T, Boman HG (1980) Insect immunity. Purification and properties of three inducible bactericidal proteins from hemolymph of immunized pupae of *Hyalophora cecropia*. Eur J Biochem 106: 7-16
14. Hultmark D, Engström Å, Bennich H, Kapur R, Boman HG (1982) Insect immunity. Isolation and structure of cecropin D and four minor antibacterial components from cecropia pupae. Eur J Biochem 127: 207-217
15. Hultmark D, Engström Å, Andersson K, Steiner H, Bennich H, Boman HG (1983) Insect immunity. Attacins, a family of antibacterial proteins from *Hyalophora cecropia*. EMBO J 2: 571-576
16. Jarosz J (1979) Simultaneous induction of protective immunity and selective synthesis of hemolymph lysozyme protein in larvae of *Galleria mellonella*. Biol Zentralbl 98: 459-471
17. Jollès J, Schoentgen F, Croizier G, Croizier L, Jollès P (1979) Insect lysozymes from three species of Lepidoptera: Their structural relatedness to the C (chicken) type lysozyme. J Mol Evol 14: 267-271
18. Kockum K, Faye I, Hofsten v. P, Lee J-Y, Xanthopoulos KG, Boman HG (1984) Insect immunity. Isolation and sequence of two cDNA clones corresponding to acidic and basic attacins from *Hyalophora cecropia*. EMBO J 3: 2071-2075
19. Kreil G, Haiml L, Suchanek G (1980) Stepwise cleavage of the pro part of promelittin by Dipeptidylpeptidase IV. Evidence for a new type of precursor – product conversion. Eur J Biochem 111: 49-58
20. Lee J-Y, Edlund T, Ny T, Faye I, Boman HG (1983) Insect immunity. Isolation of cDNA clones corresponding to attacins and immune protein P4 from *Hyalophora cecropia*. EMBO J 2: 577-581
21. Merrifield RB, Vizoli LD, Boman HG (1982) Synthesis of the antibacterial peptide cecropin A (1-33). Biochemistry 21: 5020-5031
22. Mohrig W, Messner B (1968) Immunreaktionen bei Insekten I. Lysozym als grundlegender antibakterieller Faktor im humoralen Abwehrmechanismus der Insekten. Biol Zentralbl 87: 439-470
23. Okada M, Natori S (1983) Purification and characterization of an antibacterial protein from haemolymph of *Sarcophaga peregrina* (flesh-fly) larvae. Biochem J 211: 727-734
24. Powning RF, Davidson WJ (1973) Studies on insect bacteriolytic enzymes I. Lysozyme in haemolymph of *Galleria mellonella* and *Bombyx mori*. Comp Biochem Physiol B 45: 669-681
25. Pye AD, Boman HG (1977) Insect immunity III. Purification and partial characterization of immune protein P5 from haemolymph of *Hyalophora cecropia* pupae. Infect Immun 17: 408-414
26. Qu X-m, Steiner H, Engström Å, Bennich H, Boman HG (1982) Insect immunity. Isolation and structure of cecropins B and D from pupae of the Chinese oak silk moth, *Antheraea pernyi*. Eur J Biochem 127: 219-224
27. Sidén I, Boman HG (1983) *Escherichia coli* mutants with an altered sensitivity to cecropin D. J Bacteriol 154: 170-176
28. Steiner H (1982) Secondary structure of the cecropins: Antibacterial peptides from the moth *Hyalophora cecropia*. FEBS Lett 137: 283-287
29. Steiner H, Hultmark D, Engström Å, Bennich H, Boman HG (1981) Sequence and specificity of two antibacterial proteins involved in insect immunity. Nature 292: 246-248
30. Stephens JM (1959) Immune responses of some insects to some bacterial antigens. Can J Microbiol 5: 203-228
31. Stephens JM, Marshall JH (1962) Some properties of an immune factor isolated from the blood of actively immunized wax moth larvae. Can J Microbiol 8: 719-725
32. Wigglesworth VB (1964) The life of insects. The New American Library, New York

Chapter 7 Antibacterial Molecules in Annelids

P. VALEMBOIS, P. ROCH, and M. LASSÈGUES[1]

CONTENTS

1 Introduction

Antibacterial activities were described in various annelids, but rarely the characterization of the active molecules was investigated. Six years ago, we undertook a systematic study of the antibacterial molecules present in the coelomic fluid of the earthworm (Lumbricidae) *Eisenia fetida andrei (E.f. andrei)* generally found in farm manure.

The coelomic fluid filling up the general cavity of annelids has a composition varying according to both the class of annelids and the biotope. In *E.f. andrei*, it contains about 4–6 mg ml^{-1} proteins with an osmolarity corresponding to 6.5 mg ml^{-1} NaCl. It contains also numerous free cells, some of them structurally and physiologically

1 Institut de Biologie Animale, Université de Bordeaux 1, Avenue des Facultés, 33405 Talence, France

Immunity in Invertebrates
Ed. by M. Brehélin

similar to the vertebrate leukocytes and sometimes called amoebocytes, some other containing numerous lipid inclusions, only existing in annelids and called chloragocytes. The presence of an antibacterial activity was also suspected for a long time in the so-called cocoon albumen of earthworms. The cocoon albumen is a mucous substance mainly constituted of glycoproteins and polysaccharides enclosed in a chitinous envelope. In *E.f. andrei*, each cocoon contains one to eight eggs.

In the present review, we shall report our first results concerning the activities of lysis and agglutination of various vertebrate erythrocytes. Then we shall mentioned recent investigations evidencing the existence, in *E.f. andrei*, of a large spectrum of different proteins allowing the earthworms to neutralize invading bacteria.

2 Hemolysis from *E.f. andrei* and Other Annelids

Du Pasquier and Duprat [14] were the first to evidence a hemolysis activity in the coelomic fluid of *E.f. andrei* against erythrocytes of various vertebrates, including frog, sheep (SRBC), and human. The hemolytic factor was defined as a lipoprotein inhibited by 15 min heating at 56°C. Using a modified Jerne's plaque assay, they showed that the hemolytic factor was released by the chloragocytes. Ten years after these pionneer studies, investigating all the humoral defense molecules of *E.f. andrei*, we demonstrated that the hemolytic activity of the coelomic fluid is due to a sophisticated system of molecules and that hemolytic activity also exists in the cocoon albumen in which the embryos grow.

2.1 Hemolytic System of the Coelomic Fluid

The protein content of the coelomic fluid was analyzed by polyacrylamide gel electrophoresis (PAGE) under native conditions (Fig. 1). After migration, hemolytic activity was located directly on the gel by its capacity of lysing SRBC: two clear zones of lysis were detected, corresponding to two proteins stainable by Coomassie blue. SDS-PAGE of coelomic fluid, when referred to molecular weight standard proteins, gave for these proteins a mol. wt. of about 40,000 and 45,000 [25]. A recent analysis of amino acid content of these two molecules purified by chromatofocusing (see Fig. 10) evidenced very similar compositions characterized by high quantities of glutamic acid and glycine residues.

Analytical isoelectric focusing (IEF) combined with SRBC overlay, revealed that the hemolytic activity of European *E.f. andrei* is mediated by four different proteins of pI ranging from 5.9 to 6.3 (Fig. 2). Besides the protein of pI 6.0 present in all the tested earthworms, each individual expressed one of two other active proteins. Analyzing numerous worm couples and their descendants, it appears that the three proteins of pI 5.9, 5.95, and 6.3 are coded by the same gene possessing three allelic forms [23]. The combination of these three alleles can theoretically give rise to six genetic patterns corresponding to six different families, three homozygous and three heterozygous, that we effectively evidenced (Fig. 2). Among the *E.f. andrei* population har-

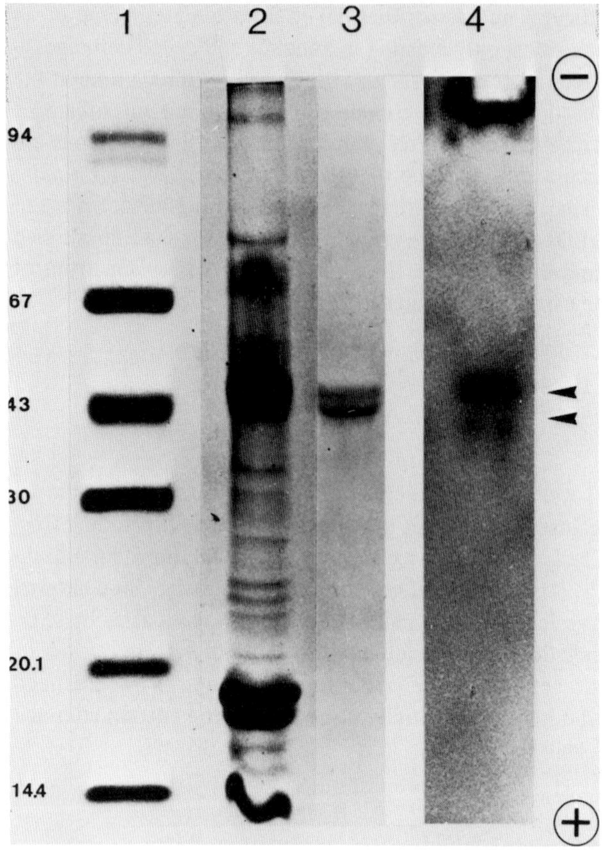

Fig. 1. Polyacrylamide gel electrophoresis (PAGE) of *E.f. andrei* coelomic fluid. 15% acrylamide gel topped by a 7% stacking gel, 1 mm thickness. The two electrode chambers were filled with the same buffer: Tris 0.05 M, boric acid 0.03 M, pH 8.80. When using denaturing conditions (*1* and *3*), 0.12% SDS was added to the gels and buffer. Migration ran 3 h at a constant current of 20 mA in a cold room. SDS gels were calibrated with α-lactalbumin (14.4 Kd), soybean trypsin inhibitor (20.1 Kd), carbonic anhydrase (30 Kd), ovalbumin 43 (Kd), bovine serum albumin (67 Kd), and phosphorylase b (94 Kd). In native conditions (*2* and *4*), migration ran 4 h at 30 mA constant current. After removing the gel sandwich, the gel No. 4 was overlaid with 1% SRBC in 0.7% agarose. The lysis (*arrows*) was observed after 20 min incubation at 22°C. Gels *1, 2,* and *3* were stained with Coomassie brilliant blue R 250. *1* Molecular weight markers under denaturing conditions; *2* coelomic fluid under native conditions; *3* SDS solubilized pellet of 55°C heated coelomic fluid under denaturing conditions; *4* Coelomic fluid under native conditions overlaid with SRBC. *Arrows* indicate the lysis spots. From Roch et al. [25]

vested in industrial lombricomposts and arising from Californian ancestors, 32% of the worms express a fourth allele never found in European worms (Fig. 2). If *E.f. andrei* is now an ubiquitous species found all over the world, it was only present in Europa before the fifteenth century. It may be hypothesized that an allele very rarely expressed in the primitive biotope has found, after migration in a new biotope, more favorable conditions to its expression.

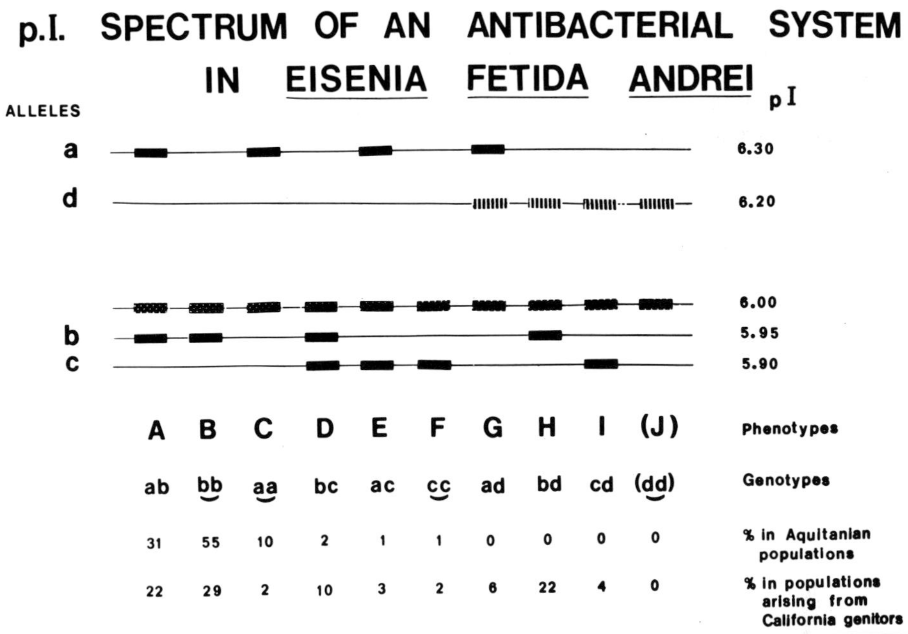

Fig. 2. Natural polymorphism of the hemolytic system of *E.f. andrei*. Schematic representation of the different hemolytic patterns as they appeared in analytical isoelectric focusing (IEF). Thin layer slab technique in a 5% polyacrylamide gel (21.5 cm x 12.5 cm and 1 mm thickness). Gels containing 4 M urea and 3.5% ampholines (LKB) were polymerized by the riboflavin-UV system. The samples, rectangular filter papers containing 10 μl, were directly applied to the gel surface, near the anode. Electrode contacts were achieved by two paper wicks impregnated with 1 M sodium hydroxide (cathode) and 1 M phosphoric acid (anode). Migration ran during 3 h at 4°C with 15 W constant power. Hemolysins were located by the property to lyse SRBC overlaid on the gel (see Fig. 1). All the proteins were then stained with Coomassie brilliant blue R 250. ▬ European isoforms (alleles a, b and c); ▬ American isoform (allele d); ▬ Common band

2.2 Hemolysin of the Cocoon Albumen

Using the same analytical methods, we showed that the cocoon albumen contained one protein of mol. wt. about 110,000 lysing the SRBC and erythrocytes of several vertebrates (Table 1). Only very small quantities of albumen can be solubilized from each cocoon, so it was not yet possible to purify enough cocoon hemolysin to determine its amino acid global composition. Also, we do not know whether this protein is coded by a polyallelic gene. Compared to coelomic fluid hemolysins, the cocoon hemolysin is characterized by different mol.wt. and physiological features: maximum titer of 1,000, unaffected by 60°C heating (Fig. 3) [19].

Table 1. Hemolysis and hemagglutination titers of *Eisenia fetida andrei* coelomic fluid and cocoon albumen on erythrocytes from different species. From Valembois et al. [32]

Erythrocytes	Sheep	Human 0	Human A	Human B	Rabbit	Rat	Badger	Chicken	Frog
Coelomic fluid									
Hemolysis[a]	42,660[c]	3,300	3,300	3,300	32,000	1,220	96	3,700	2,125
Hemagglutination[b]	500	288	288	288	557	560	16	128	4
Cocoon albumen									
Hemolysis	2,854	32	32	32	512	32	nt	500	16
Hemagglutination	64	0	0	0	16	4	nt	16	0

[a] Hemolysis tested against fresh erythrocytes
[b] Hemagglutination tested against glutaraldehyde stabilized erythrocytes
[c] Activity titers recorded as the reciprocal of the highest dilution displaying activity in the microsystem

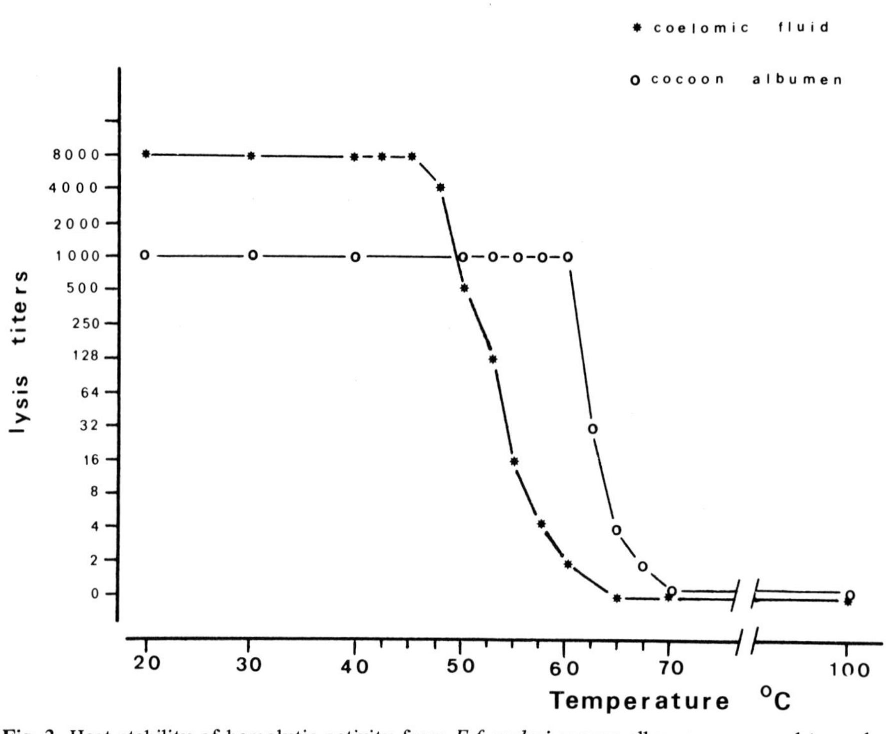

Fig. 3. Heat stability of hemolytic activity from *E.f. andrei* cocoon albumen compared to coelomic fluid. Test was applied against fresh sheep erythrocytes after 15 min heating. From Lassègues et al. [19]

2.3 Mechanism of Action

The different hemolysins of *E.f. andrei*, from the coelomic fluid as well as from the cocoon albumen, are active against most of the vertebrate erythrocytes (Table 1). The higher titer of hemolysis (42,660) was reported for SRBC [32]. Hemolytic molecules do not seem to act as enzymes since the kinetic curves of SRBC hemolysis established at 4°C and 22°C were identical. Using various preincubations with SRBC ghosts, we demonstrated that hemolysis is mediated by the binding of hemolytic molecules onto the target cells [25]. This binding occurs very rapidly (50% of the hemolysins trapped within 2 min of contact) and does not concern the Forssman antigens. Various acetylated or methylated carbohydrates inhibit the hemolysis by binding on the erythrocyte membrane, such as N-acetyl-D-glucosamine, or on the hemolysin molecules, such as α-methyl-D-mannopyranose.

2.4 Cells Synthesizing Hemolysins

The nature of the coelomic cells synthesizing the hemolysins has been investigated by hemolytic plaque assay [14], indirect fluorescence, transmission electron microscopy using peroxidase labeling [31], and scanning electron microscopy [33]. All the results indicated that both young chloragocytes, rich in granular endoplasmic reticulum, and senescent cells, sometimes called eleocytes, synthesize and release hemolysins.

The origin of cocoon hemolysin remains to be elucidated. According to preliminary observations, it seems that cocoon hemolysin would be secreted by epidermic cells of parental clitellum and incorporated to the albumen with the nutrients.

2.5 Hemolysins in Other Annelids

The coelomic fluid of other lumbricids was reported to possess sometimes a weak natural hemolytic activity [6,10,13]. The titer of hemolytic activity in earthworms other than *Eisenia* is always weak: 2 to 16 for *Lumbricus terrestris* instead of 42,660 for *E.f. andrei* with SRBC.

Several cases of hemolytic activity were recently reported in polychaetes. The coelomic fluid of *Glycera dibranchiata* lyses sheep and rabbit erythrocytes [1], chlorocruorin-free fluid drained from the crown of *Spirographis spallanzanii* lyses rat, rabbit, calf, and sheep erythrocytes [21], the coelomic fluid of *Neoamphitrite figulus* contains a hemolysin which is Ca^{2+} and Mg^{2+} dependent [12].

3 Hemagglutinins from *E.f. andrei* and Other Annelids

3.1 Hemagglutinins in the Coelomic Fluid of *E.f. andrei*

The presence of a strong hemolytic activity prevented the investigation of hemagglutination using fresh erythrocytes. Hemagglutination activity could be evidenced only when using erythrocytes stabilized by the glutaraldehyde [32]. When incubating the coelomic fluid with SRBC pretreated for 30 min with glutaraldehyde 0.05%, the agglutination titer reached 732. Such a low concentration of glutaraldehyde does not induce a spontaneous agglutination of erythrocytes, but prevents their lysis by the coelomic fluid.

Chromatographic separations performed on various gel matrices separated the coelomic fluid in 11 fractions named A to J (Fig. 4). Hemagglutination principally occurred with fractions A and B, but was also found in A′ and C. A second step of purification of active proteins from fractions A, B, and C was achieved by chromatofocusing (see Fig. 10). Assays with the protein content of the various peaks obtained by chromatofocusing, evidenced the presence of four hemagglutinins of mol. wt. 11,500, 20,000, 35,000, and 40,000 [27]. Only the molecule of mol. wt. 40,000 appears to be capable of both hemolysis and hemagglutination.

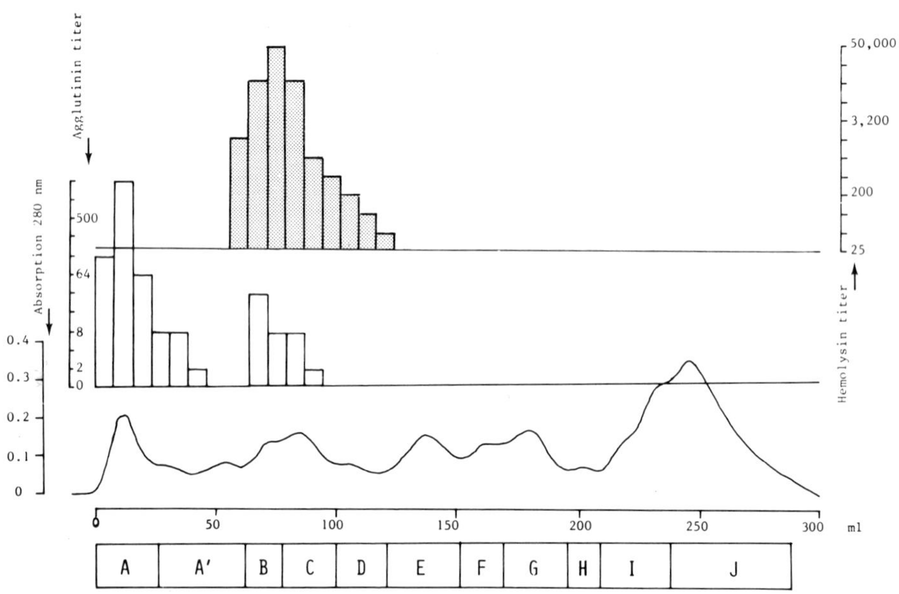

Fig. 4. Gel filtration profile of *E.f. andrei* coelomic fluid on Sephacryl S 200 (Pharmacia) matrix. A volume of 9 ml undiluted coelomic fluid, previously centrifuged at 11,000 g, was applied to the column (1.6 x 116 cm) and subjected to filtration at 15 ml h^{-1} during 24 h with 0.05 M Tris-HCl, 0.5 M NaCl, ph 8.0. Elution profile monitored at 280 nm revealed the presence of 11 major fractions indicated in block letters *A* to *J*. After elution, the various fractions were desalted and concentrated about five times. Hemolytic activity was detected against fresh erythrocytes principally in fractions *B* and *C*. Hemagglutination of glutaraldehyde stabilized erythrocytes occurred with fractions *A* and *B*, but was found also in *A′* and *C*

3.2 Hemagglutinin in the Cocoon Albumen of *E.f. andrei*

Total cocoon albumen was submitted to gel filtration on Trisacryl GF 2,000 (Pharmacia). Among the five different fractions, hemagglutination was only detected in fraction C (Fig. 5). As evidenced by gradient PAGE and SDS-PAGE, the hemagglutinin contained in fraction C has an apparent mol. wt. of 22,000 [19].

The cocoon agglutinin is distinct from the coelomic fluid agglutinins. Not only the mol. wt. is different, but also the capacity to agglutinate stabilized erythrocytes (cocoon maximum titer of 64 instead of 732 for the coelomic fluid) and the heat stability (cocoon agglutinin unaffected by a temperature of 55°C which inactivates the coelomic fluid agglutinins) (Fig. 6).

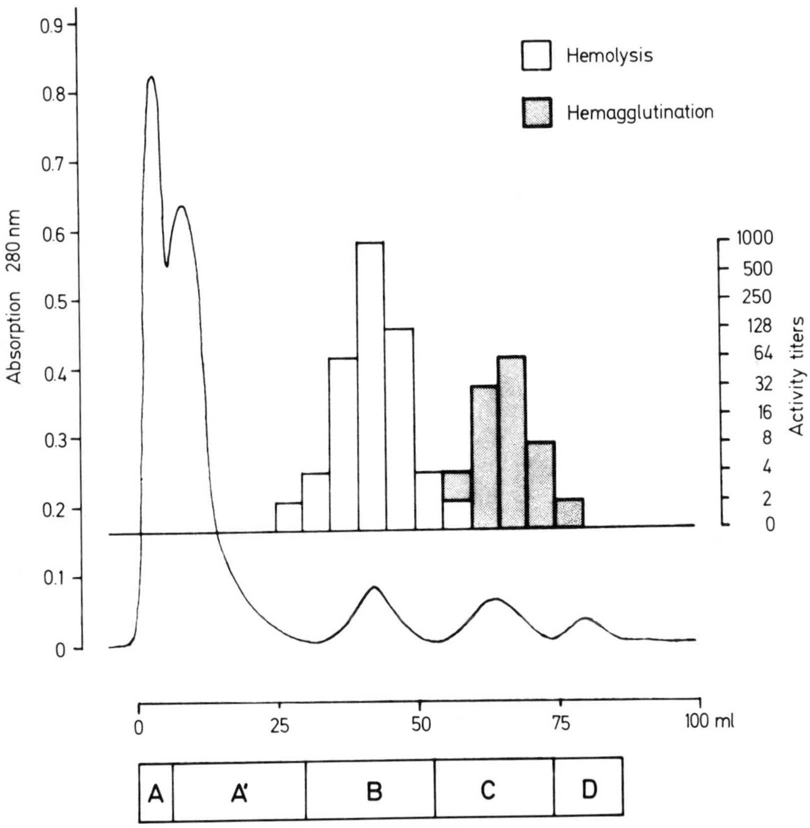

Fig. 5. Elution profile of *E.f. andrei* cocoon albumen filtrated through Trisacryl GF 2,000 (Pharmacia) matrix in a column of 1.6 x 48 cm. The sample (5 ml) consisted of pooled cocoon albumen mechanically solubilized in elution buffer (0.05 M Tris-HCl, 0.5 M NaCl, pH 8.0), eluted at 17 ml h^{-1} during 6 h. Five fractions indicated in block letters *A* to *D*, were evidenced by their absorbance monitored at 280 nm. After desalting and concentrating about five times the various fractions, hemagglutination activity can be detected in fraction *C* and hemolytic activity in fraction *B*. From Lassègues et al. [19]

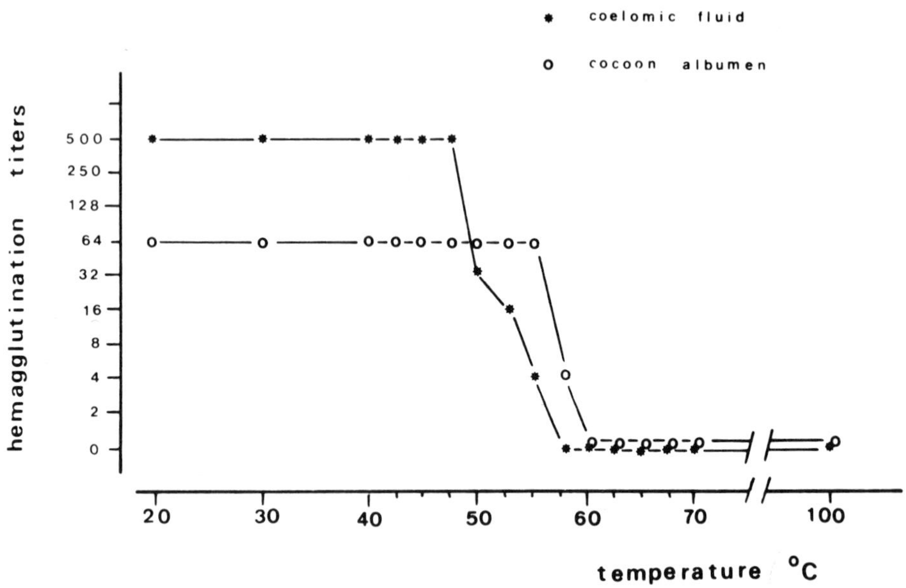

Fig. 6. Heat stability of hemagglutination activity from *E.f. andrei* cocoon albumen compared to coelomic fluid. Test against glutaraldehyde stabilized erythrocytes after 15 min heating. From Lassègues et al. [19]

3.3 Physiological Properties of Hemagglutinins

As for hemolysis, hemagglutination activity exhibited by the coelomic fluid was tested against various vertebrate erythrocytes, including sheep, human, rabbit, rat, badger, chicken, and frog (Table 1). Maximum agglutination was found with rat, rabbit, and sheep when testing coelomic fluid, and with sheep, rabbit, and chicken when testing cocoon albumen. Hemagglutination titers were always lower than hemolytic ones. Nevertheless, except for rat and frog erythrocytes, a strict parallel between both activities was noted: when for a species, one of these activities is high, so is the other [32]. Such a parallel between hemolytic and hemagglutination titers suggests that the erythrocyte determinants for both activities could be structurally related. But, if we have determined several carbohydrates capable of inhibiting hemolytic activity, we did not find any sugar capable of inhibiting hemagglutination activity. Investigations with other carbohydrates must be undertaken in order to settle whether *E.f. andrei* agglutinins are true lectins [16] as most invertebrate agglutinins [35].

3.4 Cells Synthesizing Hemagglutinins

As we did not obtain specific anti hemagglutinin antiserum, the cells responsible for hemagglutinin synthesis were indirectly investigated. A scanning electron microscopic study revealed that agglutination pictures obtained by incubating worm coelomocytes and stabilized erythrocytes together could present four different aspects (Fig. 7):

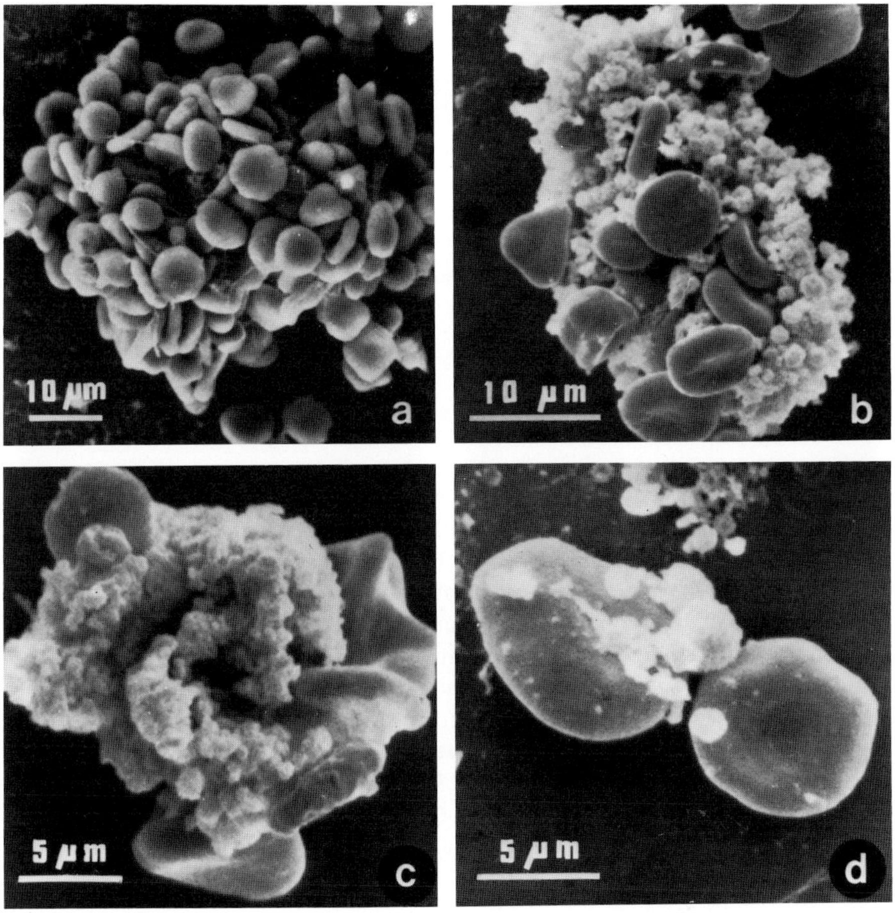

Fig. 7 a-d. Hemagglutination activity evidenced in *E.f. andrei* coelomic fluid by scanning electron microscopic observations using glutaraldehyde stabilized erythrocytes. Approximately 3 x 10^6 coelomic cells were deposited on a 18 mm round glass coverslip and mixed with 50 x 10^6 glutaraldehyde stabilized erythrocytes. Immediately after mixing, the coverslip was centrifuged for 10 min at 120 g to increase the cellular contacts and improve the cell attachment to the glass. The coverslip was then slowly dipped into the fixative mixture and maintained for 90 min at 4°C. The fixative mixture was made of 1 ml glutaraldehyde 25% commercial solution, 1.5 ml osmium tetroxide 4%, 1 ml paraformaldehyde 30%, 0.5 ml mercuric bichloride saturated solution, and 6 ml 0.1 M sodium cacodylate buffer (pH 7.5) with sucrose 3.5%. The same buffer was used to wash the coverslip after fixation. Subsequently, the cells were dehydrated by passing them through a graded series of ethanol, transferred into pentyl acetate, then into liquid CO^2. After critical point drying, the coverslip was mounted on a metal stud with silver electrical paint, coated with a thin layer of gold-palladium and examined with a Camebax scanning electron microscope (Technique from Valembois et al. [33]). **a** Mass of several hundred erythrocytes agglutinated apparently in absence of coelomocytes. **b** A large chloragocyte (about 30 μm) with a well-developed coat of chloragosomes having agglutinated about 12 erythrocytes. **c** Four to five erythrocytes agglutinated at the surface of a fragment of chloragocyte. **d** Two erythrocytes closely adhering to a small cluster of chloragosomes arising from degranulation of chloragocytes

a) masses of several hundred erythrocytes agglutinated together apparently in absence of coelomocyte;
b) rosette-like picture of 10–15 erythrocytes agglutinated at the surface of large chloragocytes (about 30 μm);
c) cluster of four to six erythrocytes adhering at the surface of a fragment of chloragocytes;
d) groups of less than five chloragosomes arising from degranulation of chloragocytes and having agglutinated one to three erythrocytes [33].
Such pictures suggest a synthesis of agglutinin by the chloragocytes, but it is also possible that the chloragocytes only fix hemagglutinin molecules synthesized by other cells.

3.5 Hemagglutinins in Other Annelids

A natural hemagglutination activity has been reported in another lumbricid *Lumbricus terrestris* [10]. This activity occurs with some mammalian erythrocytes and would be only present in certain individuals [7]. When existing, the agglutination activity has a low titer [29]. The natural hemagglutinin(s) of *L. terrestris* is unaffected by 30 min heating at 100 °C. Another difference between *Eisenia* and *Lumbricus* seems to be the cell category responsible for hemagglutinin synthesis. For Châteaureynaud-Duprat and Lestage [8] as well as for Möhrig et al. [20], the cells involved in *L. terrestris* hemagglutination seem to belong to the leukocyte (also called ameobocyte) population.

The presence of a hemagglutination activity was also reported in other classes of annelids, including polychaetes, such as *Chaetopterus variopedatus* [30], *Sabellastarte magnifica* [3], *Amphitrite ornata* [15], *Glycera dibranchiata* [1], *Neoamphitrite figulus*, and *Arenicola marina* [12] and hirudinea, such as *Hirudo officinalis* [4].

4 Characterization of Antibacterial Molecules in *E.f. andrei* and Other Annelids

4.1 Growth Inhibition Tests

Different methods have been used to investigate a possible inhibitory effect of earthworm fluids on in vitro bacteria growth. The first was the disk method, consisting of placing a disk of blotting paper impregnated with the substance to be tested on an agar plate inoculated with bacteria [26]. Such a method is simple and rapid, but mainly qualitative. So, we used then a spectrophotometric method in which the bacteria growth was followed by measuring the optical density of liquid cultures. The relationship between the number of bacteria in culture and the optical density was established by the technique of plate counts. A good precision allied to a miniaturized method is obtained now by culturing the bacteria in a liquid medium containing ^{3}H-thymidine and by following the incorporation of ^{3}H into the bacterial DNA by scintillation counting.

We successively reported bacteriostatic activity found in unseparated coelomic fluid and cocoon albumen, then in coelomic fluid fractions obtained after gel filtration and finally in peaks separated by chromatofocusing. Antibacterial activity of fractions or peaks from cocoon albumen has not yet been investigated.

4.2 Bacteriostatic Activity of Unseparated Fluids

In the past, antibacterial activities of invertebrate fluids were tested against common bacteria or against vertebrate (human) pathogenic bacteria and most of these studies failed [9]. In our case, we were interested in testing a possible activity against the telluric bacteria living in the same biotope as the earthworm *E.f. andrei*. After having isolated 23 different bacteria strains from the manure and determining some of them (Table 2), they were submitted to the action of *E.f. andrei* fluids. Only six bacteria strains were found to have their growth inhibited when cultivated in the presence of

Table 2. Recapitulative results of some biological characteristics of soil bacteria associated with the earthworm *Eisenia fetida andrei*. From Valembois et al. [31]

Species	Gram	Labeling with anti-SRBC	Bacteriostatic activity of: Coelomic fluid	Bacteriostatic activity of: Cocoon albumen	Pathogenic effect on *E.f. andrei* tissues
Bacillus megaterium	+	+	+	+	+
Arthrobacter sp.	+	+	+	+	+
Acinetobacter spl	–	–	–	–	–
Brevibacterium spl	+	–	–	–	–
Enterobacter sp.	–	–	–	–	–
Acinetobacter sp2	–	–	–	–	–
Brevibacterium sp2	+	–	–	–	–
Serratia sp.	–	+	+	+	+
Proteus vulgaris	–	–	–	–	–
Aeromonas salmonicida	–	–	–	–	–
Aeromonas spl	–	+	+	+	+
Xanthomonas sp.	–	–	–	–	–
Bacillus cereus micoïdes	+	–	–	nt[b]	–
n.i.[a]	–	–	–	–	–
Aeromonas sp2	–	–	–	–	–
n.i.	–	–	–	–	–
n.i.	+	–	–	nt	–
n.i.	+	–	–	nt	–
n.i.	–	–	–	nt	–
n.i.	–	–	–	nt	–
n.i.	–	–	–	nt	–
Achromobacter sp.	–	+	+	nt	+
Aeromonas hydrophila	–	+	+	+	+

[a] Not identified
[b] Not tested

coelomic fluid or cocoon albumen. Inoculation of a high dose of these six strains directly into the body cavity of the worms caused the death of the animals within 36 h [18]. Quantitatively similar inoculations, but of bacteria strains insensible to *E.f andrei* fluids, were never lethal for the worms. It must be noted that the six bacteria strains sensible to *E.f. andrei* fluids and found to be pathogenic for the worms expressed one (or several) common antigen(s) with vertebrate erythrocytes as revealed by their labeling with an anti-SRBC antiserum (Table 2). This common heterophilic antigen was demonstrated to be distinct from the Forssman antigen [25]. It must be also pointed out that the bacteria strains sensible to *E.f. andrei* fluids were indistinctly Gram positve, such as *Bacillus megaterium*, or Gram negative, such as *Aeromonas hydrophila*.

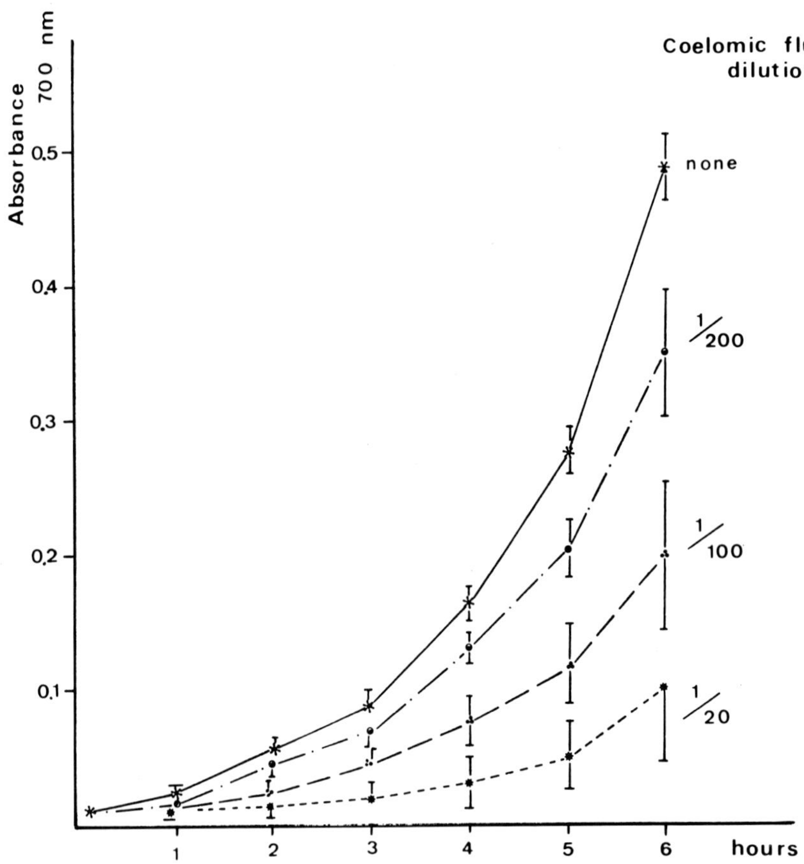

Fig. 8. Inhibitory effect of unseparated coelomic fluid on the in vitro growth of *B. megaterium*. Log-phase growing bacteria were centrifuged and resuspended in nutritive broth (pancreatic peptone 10 g, meat extract 5 g, NaCl 5 g, distilled water 1 l) to obtain a density of 2×10^5 bacteria/ml, as determined by plate counts. 650 μl of bacteria suspension were distributed in sterile plastic spectrophotometric microcells, receiving 1/20, 1/100, or 1/200 final dilution of coelomic fluid. The microcells were incubated at 21°C under 200 rpm orbital shaking and their absorbance at 700 nm periodically recorded

The bacteriostatic effect of both coelomic fluid and cocoon albumen was strongly reduced by 15 min heating at 55°C. Based on similarity of heat sensitivity, we hypothesized in 1982 [31] that at least one of the bacteriostatic factors belongs to the hemolytic system. This has now been demonstrated after purification and biochemical analysis of the various active molecules. Antibacterial activity of both unseparated coelomic fluid and cocoon albumen was dose dependent and still active at a dilution of 1/200 (Fig. 8).

Part of these studies was reinvestigated by Çotuk and Dales [11] with the same conclusions, but sometimes slightly different results. In fact, we still noted that the activity of *E.f. andrei* fluids can be different if tested against different strains of the same bacteria species.

4.3 Bacteriostatic Effect of Coelomic Fluid Fractions

After filtration through a chromatographic column (Fig. 4), the various coelomic fluid fractions were desalted and their protein content determined by Lowry's method. In order to allow comparison between the different fractions, adjustments were made to obtain equal protein dilutions in each fraction (0.65–0.70 mg mg^{-1}). Two bacteria strains (*B. megaterium* and *A. hydrophila*) previously demonstrated to be sensible to *E.f. andrei* coelomic fluid [26], have been submitted to the activity of the different gel filtration fractions and have shown slightly different sensitivity.

Concerning *B. megaterium*, tests performed with gel filtration fractions revealed that fraction B was the most powerful with 30% inhibition when measured by optical density (Fig. 9). Fractions A, C, and J expressed lower activity with 17, 15, and 13% inhibition, respectively. No significant inhibition of bacterial growth was found with other fractions.

With *A. hydrophila*, as with *B. megaterium*, a significant bacteriostatic effect was found with fractions A, B, and C, but fraction A′ also expressed an important inhibitory activity. The main difference between *A. hydrophila* and *B. megaterium* probably consists in the high sensitivity of *A. hydrophila* to small molecules contained in fractions G to J. These four fractions provoked a growth inhibition of at least 60% (Fig. 9). We do not know whether this activity is due to the same molecule present in the different fractions or to several different molecules. Using a mol. wt. calibration kit, the fractions G to J were determined as containing only molecules of mol. wt. lower than 5,000. The matrices we used for gel filtration are inefficient to correctly separate such mol. wt. molecules. We intend to use more appropriate chromatographic separations (i.e., ion exchanger) to determine the number and the biochemical features of the low mol. wt. antibacterial molecule(s).

4.4 Bacteriostatic Effect of Chromatofocusing Peaks

Chromatofocusing of fraction A from gel filtration separated eight major peaks (Fig. 10). One of them, with a mol. wt. of about 175,000, seemed to possess a strong bacteriostatic activity (Fig. 9). This molecule, perhaps also present in fraction A′, is a po-

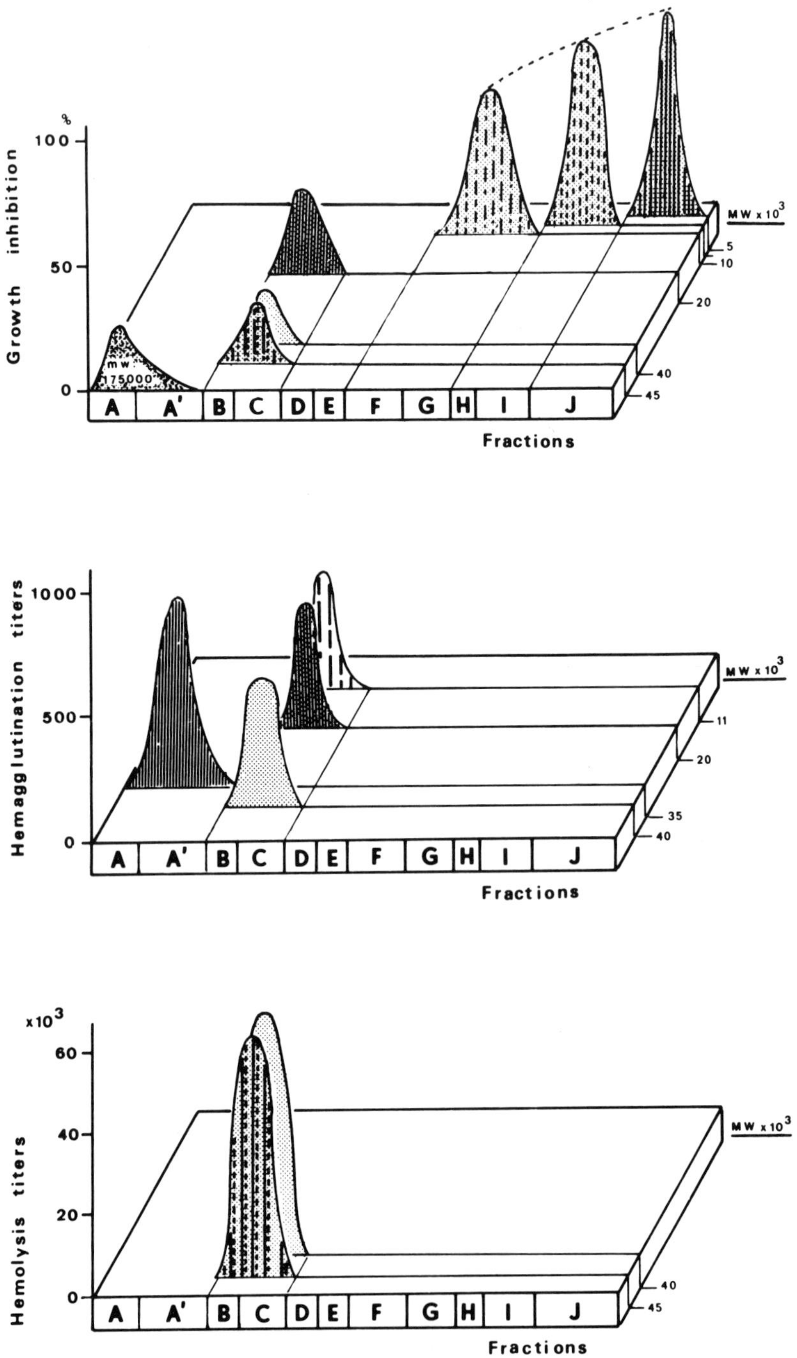

Fig. 9. Recapitulative diagrams of bacteriostatic, hemolytic, and hemagglutination activities naturally expressed by the coelomic fluid of *E.f. andrei*. Note that most of the active proteins possess several activities

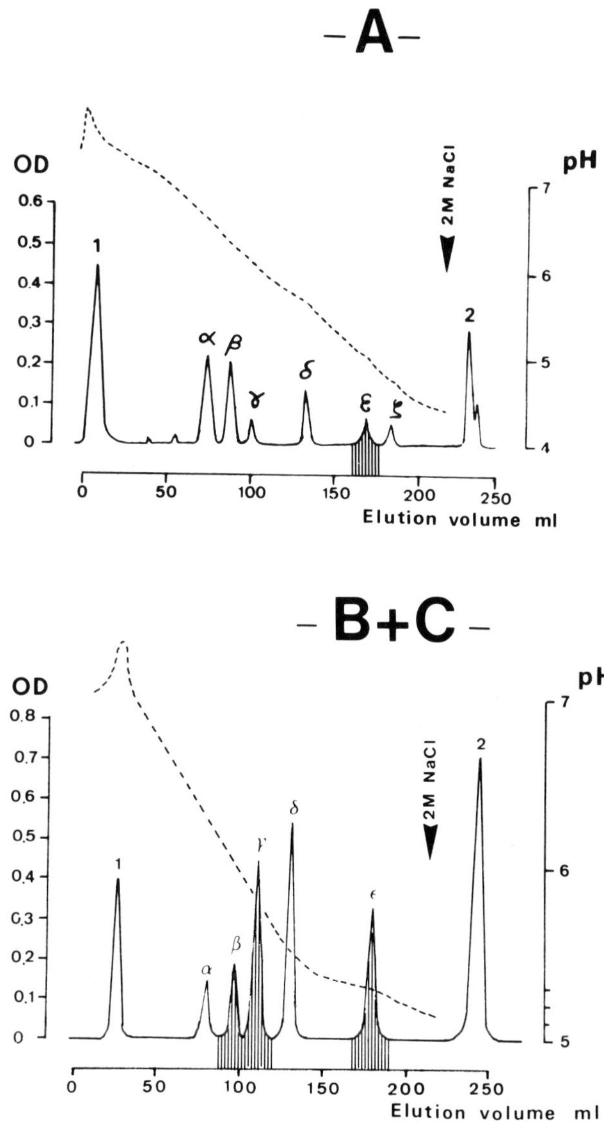

Fig. 10. Chromatofocusing patterns of gel filtration fractions *A* and *B-C* (see Fig. 4) of *E.f. andrei* coelomic fluid. According to the pH gradient of elution, the gel filtration fractions were equilibrated with start buffer (0.25 M imidazole-HCl, pH 7.4) and concentrated to a volume of 10 ml by ultrafiltration on Amicon PM 10 membrane. Chromatofocusing was performed in a column of 0.9 x 27 cm filled with 17 ml of PBE 94 gel (Pharmacia) previously equilibrated with start buffer. After 1 h packing with start buffer (60 ml h^{-1}), 1 ml of Sephadex G 50 (Pharmacia) was layered on top of the PBE 94 gel. The column was then equilibrated with start buffer (60 ml h^{-1}, 5 h). The sample (10 ml in start buffer) was applied by first running 5 ml (25 ml h^{-1}) of eluent PB 74 (Pharmacia) diluted 1:8 with water and adjusted at pH 4.0 (*A*) or 5.0 (*B-C*) with 1 M HCl, followed by applying the sample and then switching back to the eluent (25 ml h^{-1}, 8 h). The pH (*dotted line*) and protein (*solid line* monitored at 280 nm) profiles were recorded simultaneously. The localization of the four antibacterial proteins is indicated by *hatchings*

lymeric structure consisting of five identical monomers of mol. wt. 35,000. The monomer was previously reported as corresponding to a hemagglutinin [27].

The poorly marked separation of B and C gel filtration fractions provokes an overlapping between the two chromatofocusing elution profiles. So, B and C fractions have been pooled before separation in chromatofocusing (Fig. 10). A significant bacteriostatic activity was noted with three (out of seven) different peaks, mol. wt. 20,000, 40,000, and 45,000 (Fig. 9). These molecules are also capable of either hemolytic or hemagglutination activities. Analyzing the activity of only the high mol. wt. proteins of the coelomic fluid leads to evidence of at least four different antibacterial molecules.

In order to correlate the antibacterial activities of the various fractions and peaks with that of unseparated coelomic fluid, the protein content of each fraction or peak was adjusted according to the elution profiles (Figs. 4 and 10). The theoretical total of activity obtained by addition of the activities expressed by the various fractions or peaks was always lower than the activity obtained with unseparated coelomic fluid. Either a part of the activity was lost during chromatographic separations or a cooperation between several fluid components exists. Numerous attempts to recover a maximum bacteriostatic activity by using pools of several fractions were unsuccessful. The same phenomenon was also reported by Anderson [1] in *Glycera dibranchiata*.

Investigations with radioactive precursors of nuclei acids and proteins are in progress to determine the bacterial metabolic pathway affected by each antibacterial molecule.

4.5 Other Antibacterial Factors Found in Annelids

The presence of lysozyme-like substances in several annelids was reported 25 years ago. Vertebrate lysozyme, isolated from tissue secretions or egg white, is a light globular polypeptide acting by splitting the bond between N-acetyl-muramic acid and N-acetyl-glucosamine of bacteria peptidoglycans. N-acetyl-muramidases physiologically similar to lysozyme, but structurally very different, exist in various invertebrates. Some factors which are active both on Gram positive and Gram negative bacteria were sometimes called abusively lysozymes.

Jollès and Zuili [17] evidenced in the polychaete *Nephthys hombergi* a factor inhibiting the growth of *Micrococcus lysodeikticus*. As the egg white lysozyme, this factor is characterized by a relative heat stability at low pH. Its mol. wt. and amino acid composition are rather different from egg white lysozyme [22].

A bacteriolytic activity assimilated to a lysozyme activity was reported by Schubert and Messner [28] in nine annelids, including polychaetes, oligochaetes and hirudinea. Various tissue extracts or body fluids were found active against Gram positive bacteria, such as *Micrococcus lysodeikticus* and *Bacillus subtilis*, but also against Gram negative bacteria, such as *Aeromonas* sp. Çotuk and Dales [11] described in *E. fetida* a lysozyme-like substance with a mol. wt. of 15,000. This substance, active against both Gram positive and negative bacteria, expressed its maximum activity at pH 6.2 and presented a slightly enhanced activity following an injection of formalized *M. lysodeikticus*.

An antibacterial factor, different from lysozyme, was evidenced by Anderson [1] in the coelomic fluid of the polychaete *Glycera dibranchiata*. This factor was demonstrated to be bacteriostatic for Gram negative bacteria, including *Serratia marcescens, Pseudomonas aeruginosa*, and some *E. coli* strains. The bacteriostatic activity of *G. dibranchiata* coelomic fluid is mediated by a heat labile protein with a mol. wt. higher than 10,000 [2]. The molecule-bacteria interaction involves two consecutive steps [5]: an initial stage of binding to the bacterial wall requiring divalent cations, followed by a second stage of killing the bacteria.

5 Conclusion

It is now admitted that lumbricids are able to reject a second graft, genetically related to the first one, in an accelerated fashion. So, lumbricids can mediate anamnestic reactions of cellular immunity. Concerning a possible induction of the humoral defenses, the question gives rise to much controversy. According to Dales [12] and to Möhrig et al. [20], there is no evidence for an enhanced agglutination response in annelids following injection of eryhtrocytes. According to Châteaureynaud-Duprat and Izoard [7] and Stein et al. [29], it is possible to increase the naturally occurring hemagglutination activity of *Lumbricus terrestris* by challenging the worms. The conditions for obtaining an enhanced response were well established by Stein et al. [29]. These authors showed that the agglutination titer increased from 14 to 1,216, 24 h after a single injection of erythrocytes. Injection of different carbohydrates, amino acids, nucleotides, proteins etc., also increased the agglutination titer. Simultaneously with the enhanced hemagglutination activity, an increased concentration of the different coelomic fluid proteins was noted [34]. So, a first injection of various antigens or haptens indistinctly provokes an increased activity presenting no specificity. We did not investigate the ability of inducing enhanced antibacterial responses in *E.f. andrei*, but we noticed that animals bred in the laboratory under artificial conditions [24] partially lost their antibacterial activity. Our opinion is that a first contact with pathogenic bacteria is very likely necessary to induce an efficient antibacterial defense.

The humoral antibacterial molecules evidenced in *E.f. andrei* and other annelids do not express a specificity of recognition analogous to that of vertebrate immunoglobulins. They probably recognize simple haptenic groups present at the surface of numerous cells and bacteria and sometimes called heterophilic antigens. This opinion is in agreement with the possibility of inhibiting hemolysis [25] or hemagglutination [12] with various carbohydrates and particularly with N-acetylated ones. The fact that the carbohydrates inhibiting hemolysis have no effect on hemagglutination and that the coelomic fluid fractions active on *A. hydrophila* are not necessarily active on *B. megaterium*, demonstrate that a species like *E.f. andrei* is able to recognize a large spectrum of heterophilic antigens. This recognition could be mediated by a variety of lectin-like molecules. Some of these lectin-like molecules are coded by polyallelic genes (i.e., the hemolytic system of *E.f. andrei*) increasing the spectrum of available antibacterial factors. So, the antibacterial defense system of annelids appears to be mediated not by one or two "nonspecific factors", but rather by a polymorphic system of molecules, each of them "specific" for one heterophilic antigen.

References

1. Anderson RS (1980) Hemolysins and hemagglutinins in the coelomic fluid of a Polychaete Annelid, *Glycera dibranchiata*. Biol Bull (Woods Hole) 159: 259-268
2. Anderson RS, Chain BM (1982) Antibacterial activity in the coelomic fluid of a marine Annelid, *Glycera dibranchiata*. J Invertebr Pathol 40: 320-326
3. Brown R. Almodovar LR, Bhatia HM, Boyd WC (1968) Blood group specific agglutinins in invertebrates. J Immunol 100: 214-216

4. Brundshuh G (1966) Agglutinierende Antikörper gegen Humanerythrozyten bei *Hirudo officinalis*. Z Aerztl Forbild 60: 196-206
5. Chain BM, Anderson RS (1983) Antibacterial activity of the coelomic fluid of the Polychaete, *Glycera dibranchiata* I. The kinetics of the bactericidal reaction. Biol Bull (Woods Hole) 164: 28-40
6. Châteaureynaud-Duprat P, Izoard F (1973) Etude des mécanismes de défense chez *Lumbricus terrestris*. C R Acad Sci Paris 276: 2859-2862
7. Châteaureynaud-Duprat P, Izoard F (1977) Etude comparée in vitro des réactions de défense antigreffe ches 2 genres de Lombriciens: *Eisenia* et *Lumbricus*. C R Acad Sci Paris 284: 2581-2584
8. Châteaureynaud-Duprat P, Lestage J (1980) Etude in vitro des diverses fonctions des leucocytes de *Lumbricus terrestris* dans les réactions immunitaires. C R Acad Sci Paris 291: 889-892
9. Cooper EL, Acton RT, Weinheimer PF, Evans EE (1969) Lack of bactericidal response in the earthworm *Lumbricus terrestris* after immunization with bacterial antigens. J Invertebr Pathol 14: 402-406
10. Cooper EL, Lemmi CAE, Moore TC (1974) Agglutinins and cellular immunity in earthworms. Ann NY Acad Sci 234: 34-49
11. Çotuk A, Dales RP (1984) The effect of the coelomic fluid of the earthworm *Eisenia fetida* (Sav) on certain bacteria and the role of the coelomocytes in internal defense. Comp Biochem Physiol 78A: 271-275
12. Dales RP (1982) Haemagglutinins and haemolysins in the body fluids of *Neoamphitrite figulus* and *Arenicola marina* (Annelida, Polychaeta). Comp Biochem Physiol 73A: 663-667
13. Du Pasquier L (1971) Etude comparée d'un facteur cytolytique humoral chez une larve d'amphibien et chez un Oligochète. Arch Zool Exp Gen 112: 81-87
14. Du Pasquier L, Duprat P (1968) Aspects humoraux et cellulaires d'une immunité naturelle non spécifique chez l'Oligochète *Eisenia fetida* (Lumbricidae). C R Acad Sci Paris 266: 538-541
15. Garte SJ, Russel CS (1976) Isolation and characterization of a hemagglutinin from *Amphitrite ornata*, a polychaetous annelid. Biochim Biophys Acta 439: 368-379
16. Goldstein IJ, Hughes RC, Monsigny M, Osawa T, Sharon N (1980) What should be called a lectin? Nature 285: 66
17. Jollès P, Zuili S (1960) Purification et étude comparée de nouveaux lysozymes; extraits du poumon de poule et de *Nephthys hombergi*. Biochim Biophys Acta 39: 212-217
18. Lassègues M, Roch P, Valembois P, Davant N (1981) Action de quelques souches bactériennes telluriques sur le Lombricien *Eisenia fetida andrei*. C R Acad Sci Paris (série III) 292: 731-734
19. Lassègues M, Roch P, Cadoret MA, Valembois P (1984) Mise en évidence de protéines hémolytiques et hémagglutinantes spécifiques de l'albumen des cocons du Lombicien *Eisenia fetida andrei*. C R Acad Sci Paris (série III) 299: 691-696
20. Möhrig W, Kauschke E, Ehlers M (1984) Rosette formation of the coelomocytes of the earthworm *Lumbricus terrestris* (L) with sheep erythrocytes. Dev Comp Immunol 8: 471-476
21. Parrinello N, Rindone D (1981) Studies on the natural hemolytic system on the annelid worm *Spirographis spallanzanii* Viniani. Dev Comp Immunol 5: 33-42
22. Périn JP, Jollès P (1972) The lysozyme from *Nephthys hombergi* (Annelid). Biochim Biophys Acta 263: 683-689
23. Roch P (1979a) Protein analysis of earthworm coelomic fluid I: Polymorphic system of the natual hemolysins of *Eisenia fetida andrei*. Dev Comp Immunol 3: 599-608
24. Roch P (1979b) Leukocyte DNA-synthesis in grafted lumbricids: an approach to study histocompatibility in invertebrates. Dev Comp Immunol 3: 417-428
25. Roch P, Valembois P, Davant N, Lassègues M (1981) Protein analysis of earthworm coelomic fluid II: Isolation and biochemical characterization of the *Eisenia fetida andrei* factor (EFAF). Comp Biochem Physiol 69B: 829-836
26. Roch P, Valembois P, Lassègues M, Cassand P (1981) Protective activity of the coelomic fluid of a lumbricid *Eisenia fetida*. Dev Comp Immunol 5: 81-86
27. Roch P, Davant N, Lassègues M (1984) Isolation of agglutinins from lysins in earthworm coelomic fluid by gel filtration followed by chromatofocusing. J Chromatogr 290: 231-235

28. Schubert I, Messner B (1971) Untersuchungen über das Vorkommen von Lysozym bei Anneliden. Zool Jahrb Physiol 76: 36-50
29. Stein EA, Wojdani A, Cooper EL (1982) Agglutinins in the earthworm *Lumbricus terrestris*: naturally occurring and induced. Dev Comp Immunol 6: 407-421
30. Tyler A (1946) Natural heteroagglutinins in the body fluids and seminal fluids of various invertebrates. Biol Bull (Woods Hole) 90: 213-219
31. Valembois P, Roch P, Lassègues M, Cassand P (1982) Antibacterial activity of the hemolytic system from the earthworm *Eisenia fetida andrei*. J Invertebr Pathol 40: 21-27
32. Valembois P, Roch P, Lassègues M (1984) Simultaneous existence of hemolysins and hemagglutinins in the coelomic fluid and in the cocoon albumen of the earthworm *Eisenia fetida andrei*. Comp Biochem Physiol 78A: 141-145
33. Valembois P, Lassègues M, Roch P, Vaillier J (1985) Scanning electron microscopic study of the involvement of coelomic cells in earthworm antibacterial defense. Cell Tissue Res 240: 479-484
34. Wojdani A, Stein EA, Lemmi CA, Cooper EL (1982) Agglutinins and proteins in the earthworm *Lumbricus terrestris* before and after injection of erythrocytes, carbohydrates and other materials. Dev Comp Immunol 6: 613-624
35. Yeaton RW (1981) Invertebrate lectins II: Diversity of specificity, biological synthesis and function in recognition. Dev Comp Immunol 5: 535-545

Chapter 8 Invertebrate Lectins: Biochemical Heterogeneity as a Possible Key to Their Biological Function

J. A. OLAFSEN[1]

CONTENTS

1 Introduction

The molecular basis for discrimination between self and non-self, a crucial distinction at all levels of organization of living matter, remains one of the intriguing puzzles of immunobiology. In order to survive, all animals have evolved some kind of response to recognize and eliminate non-self without destroying self. The first cell to appear that was specialized for use in host defense was the mononuclear phagocytic cell. The amoebocytes, coelomocytes, or hemocytes of invertebrates are believed to be ancestors of the monocyte/macrophage system found in higher vertebrates and man. Phagocytosis is a fundamental attribute of many cells, and is particularly important for a vast number of invertebrates, since the process is related to nutrition and excretion as well as to defense[25].

Many invertebrates, particularly marine forms, live in areas rich in organic nutrients with an abundant microflora that, apart from providing food particles, also act as a potential source of infection [19]. Although invertebrates lack inducible immunoglobulins or a complement system, they still normally manage to keep their internal body fluids sterile [4], and it seems beyond doubt that they have evolved effective means to distinguish between self and non-self.

1 Institute of Fisheries, University of Tromsø, p.o. box 3083 Guleng, N-9001 Tromsø, Norway

Immunity in Invertebrates
Ed. by M. Brehélin

The role of hemocytes in digestion, particularly in molluscs, has been extensively discussed, but it is now generally agreed that intracellular digestion is more important than extracellular [12,13]. This complex cellular function is carried out by both fixed and circulating hemocytes, and is facilitated by the wandering nature of these cells. All these hemocytes appear to be phagocytic [26]. In molluscs the hemocytes are distributed throughout the circulatory system and because this system is an open one, they are also found within the tissues. There is also a two-way traffic of hemocytes across the epithelial lining of the alimentary canal [21].

Hydrophobicity and charge interaction are important factors in phagocytosis [91]. Bacteria more hydrophobic or negatively charged than the phagocyte are readily taken up, while those more hydrophilic are not easily engulfed. However, more precise mechanisms must be involved in the specific recognition of physiologically important particles. Glycoproteins are involved in binding bacteria to macrophages, and it is conceivable that glycoprotein recognition determinants also function in invertebrate phagocytosis (for a recent discussion, see [77]). Indeed, it seems that surface glycoproteins are involved at all levels of cellular interactions, and that carbohydrate recognition is a phylogenetically ancient binding principle represented throughout the biological world [65,77].

Lectins are carbohydrate binding, nonimmunoglobulin proteins that agglutinate cells and/or precipitate glycoconjugates through interaction with glycoproteins or glycolipids [7]. Although the carbohydrate binding specificity is unanimous for binding by lectins, secondary forces such as hydrophobicity and charge interaction, possibly operating through multivalent binding, may also stabilize the bond. Lectins are found throughout the biological world, and have been described from viruses, bacteria, yeasts, plants, invertebrates, and vertebrates [34]. More recently lectins have been identified as membrane components of various cells [5]. Since the now classical review by Sharon and Lis [78] in 1972, a number of books and review articles have appeared that describe the occurrence, structure, specificity, and possible function of lectins [34,35, 61,76], and their interaction with cell surfaces [55]. However, comparatively little effort has been directed toward understanding the relationship between the molecular architecture and the biological function of these molecules. Because of their ability to agglutinate cells, lectins have lent themselves easily to speculation as to whether they are primitive recognition molecules, even though they bear little or no relationship to vertebrate immunoglobulins. In fact, recent evidence points to lectins being a heterogeneous class of proteins with little or no structural or functional relationship, and it is still an open question whether their in vitro ability to agglutinate cells has any universal biological significance. Apart from taking part in self/non-self recognition, it has been proposed that lectins may effect signals across biological membranes, partake in the organization of macromolecules or multienzyme complexes or even promote transport of calcium or sugars.

The term opsonin is used to represent any agent in serum which acts on particles to increase their ingestion by hemocytes. A number of reports on the opsonic properties of hemolymph factors from different invertebrates are now available [2,52,57-59, 63,64,70,84]. However, most are concerned with crude or partly purified hemolymph factors, and only rarely have purified lectins been used. No phagocytosis of erythrocytes by lobster hemocytes was observed in the absence of purified lobster lectins, whereas in their presence uptake was occasionally stimulated [37]. The increased in vitro phagocytosis of lectin-coated bacteria by oyster hemocytes [40], clearly demonstrated that purified oyster lectins promoted phagocytosis of bacteria by their indigenous hemocytes. Opsonizing properties of an isolated *Mytilus* lectin were also reported [67], and the presence of identical lectins in the hemolymph and on the surface of hemocytes has been demonstrated in a number of animals [67,82,90,96], indicating that both cell-bound and circulating factors are involved in non-self recognition. Although an absolute requirement for opsonic factors is sometimes reported [37,63], this is unusual, since uptake of erythrocytes, bacteria, or particles by invertebrate hemocytes also occurs independently of serum factors [64,73,75,87]. Moreover, hemocytes may provide some degree of specificity, as demonstrated in the selective phagocytosis of Gram positive or Gram negative bacteria effected by different subpopulations of *Arenicola marina* hemocytes [24].

A number of lectins are polymeric molecules built from one type of subunit with a single detectable binding site. No matter how sophisticated the carbohydrate binding specificity might be, opsonization would be a chance event without a second cytophilic binding site, which should normally be masked or deactivated to avoid reaction with self. With the number of antigens that invertebrates encounter, a model based on broad specificity or heterogeneity, thus, has certain advantages. Whereas a number of homogenous lectins have been described, an increasing degree of heterogeneity seems to emerge; as evident from the presence in the hemolymph of multiple lectins, isolectins, or lectins with heterologous binding sites. Only this last group would have a molecular architecture implying both self and non-self recognition elements.

Jenkin and Hardy [44] proposed a hypothesis for "the generation of diversity" working with multispecific lectins from the crayfish. Agglutinating activity to sheep erythrocytes was removed by absorption with these cells, but could subsequently be restored by treating the hemolymph first with EDTA and then with Tris/calcium buffer. It was inferred that this treatment caused the random association of monomers into new functional lectins. A theoretical, but less conceivable model was proposed by Parish [56], implying the random association of glycosyltransferases (recognizing non-self) with the addition of one self determinant. However, experimental evidence in support of this model is still awaited.

Invertebrate immunity has been extensively reviewed [4,5,14,17,25,44,48,65,85] as have both the occurrence and properties of invertebrate lectins [65,99,100], but discussions with a functional approach based on both biochemical and biological data are rare. This paper will focus mainly on heteroagglutinins, with special reference to molluscs, and in particular concentrate on how the complex biological function relating to these molecules might possibly be inferred from their molecular architecture. For further details, the reader is referred to the comprehensive review of Ey and Jenkin [20] with additional help from a very useful Table provided by Fries [31].

2 A Survey of Invertebrate Lectins with Emphasis on Heterogeneity

2.1 Primitive Metazoa

Sponge lectins are involved in aggregation phenomena allowing cell to cell contact in sponge tissues. The lectins from *Axinella polypoides* [9] were low molecular weight proteins (M_r 21,000 and 15,000, respectively) with only minor differences in specificity, both reacting with oligosaccharides containing β-linked D-galactose. Lectins from *Aaptos papillata* [10] were separated into three fractions with different isoelectric points, and from SDS-PAGE it was demonstrated that Aaptos II and III lectins consisted of a single subunit (M_r 16,000), whereas Aaptos I yielded two subunits (M_r 21,000 and 12,000). *Halichondria panicea* lectin [54] was a M_r 78,000 protein consisting of four identical subunits (M_r 21,000). It was put forward as the likely basis for a symbiotic relationship between the sponge and a bacterium.

2.2 Mollusca

The molluscs have probably been investigated for the presence of lectins more than any other group of invertebrates, and some species studied in considerable detail. Prowse and Tait [63] showed that hemolymph promoted in vitro phagocytosis of yeast cells and sheep erythrocytes by *Helix aspersa* amoebocytes. Renwrantz and his colleagues [51,66,68] demonstrated increased clearance of human (A and B) and rabbit erythrocytes from the circulation when preincubated with hemolymph. A secondary injection was less effectively cleared unless the erythrocytes were pretreated with hemolymph. Evidence was presented to indicate that *Helix* lectin and opsonin were the same.

The blood group A specific lectin from *Helix pomatia* (anti-A_{HP}) purified from the albumin gland of adult snails [38] was a glycoprotein with M_r 79,000 [39] consisting of six identical subunits arranged as disulfide bonded dimers. Each subunit contained one intrachain disulfide bond and a carbohydrate-binding site. Optical rotatory dispersion and circular dichroic spectra [38] indicated predominant β-structure. An aggregated, "superagglutinating", form of the molecule has been reported [88]. Later *Helix pomatia* lectins [98] were fractionated into (at least) 12 isolectins. Each intact lectin had a M_r of about 80,000, and it was suggested that heterogeneity of *H. pomatia* lectin could result from synthesis of a number of different lectin molecules, or combinations of different polypeptide chains to form molecules of different structures. A mucus lectin from *Helix aspersa* [29] was heterogeneous, giving three protein bands by PAGE, and was resolved in (at least) ten bands by isoelectric focusing. Heterogeneity of *Biomphalaria glabrata* lectin was also demonstrated [11]; lectin II (M_r 350,000) agglutinated human erythrocytes (A and B), and was heterogeneous in discontinuous PAGE. Lectin I (M_r 67,000), homogeneous in PAGE and electrofocusing, was composed of three M_r 17,000 subunits and one M_r 15,000 subunit, and was apparently specific for NANA.

A sialic acid-specific lectin (LFA) purified from the slug *Limax flavus* [53] appeared homogeneous by SDS-PAGE, and was revealed as a M_r 44,000 dimer composed of two equally size subunits. The lectin dissociated at pH 2.0 and tended to associate into larger molecular aggregates at concentrations above 1 mg ml^{-1} at pH 7.5, when it behaved as a rapidly associating/dissociating system as evidenced by sedimentation velocity and near-UV circular dichroic spectra. LFA appeared to be highly specific for sialic acid, but BSM was also a potent inhibitor.

Some bivalve lectins have been studied in detail. Tripp [84,86] examined hemolymph opsonic factors and agglutinins in *Crassostrea virginica* and high titers were observed with human and horse erythrocytes. Acton and colleagues [1] purified a lectin for sheep erythrocytes from *C. virginica* hemolymph. The lectin activity was associated with a heterogeneous group of rapidly sedimenting (polymeric) molecules, apparently consisting of a single polypeptide subunit (M_r 20,000 from SDS-PAGE) which contained 13% carbohydrate, and required divalent cations for agglutination. Hardy, Grant, and Fletcher [41] described agglutinating activity against a variety of erythrocytes in the hemolymph of the Pacific oyster *Crassostrea gigas*. Activity against both human and horse erythrocytes revealed separate entities, named Gigalin H (human) and Gigalin E (equine), respectively, which were purified by affinity chromatography on BSM-Sepharose 4B. Gigalin H was specific for NANA residues, but it was notable that the inhibitory effect of BSM, expressed in terms of sialic acid, was 16,000 times greater than that of free NANA. It was, thus, suggested that Gigalin H had multiple binding sites with affinity for sialic acid residues.

In vivo exposure of *C. gigas* to bacteria in seawater brought about increased lectin activity of oyster hemolymph, suggesting the possible involvement of these molecules in a defense reaction [40]. Later we demonstrated the effect of in vivo exposure of oysters to the pathogen *Vibrio anguillarum* NCMB 6 at different time intervals, temperature, and seasons (Olafsen J.A., Fletcher T.C., Grant P.T., in preparation), and though there was great biological variation, significantly enhanced lectin activity resulted from this challenge. The response was more marked for Gigalin E than for Gigalin H, and it was only transient, i.e., the activity returned to normal after about a week; by then the water was usually cleared of bacteria. The response was not due to cell damage caused by the live pathogen, since dead bacteria also induced a similar response (albeit lower). Both hemolymph and purified lectins acted as opsonins, i.e., increased phagocytosis of bacteria (*V. anguillarum* NCMB 6 and *Escherichia coli* K235) by oyster hemocytes in vitro [40].

Although Gigalin E had no affinity for sialic acid or BSM, separation of Gigalin H and Gigalin E by affinity chromatography was impossible [63]. Purification following either absorption of all detectable Gigalin E with glutaraldehyde-treated horse erythrocytes, or by rechromatography of the nonbinding fraction from the BSM-Sepharose column (devoid of Gigalin H activity) always resulted in a mixture of the two activities following elution and dialysis; thus, suggesting that purified Gigalin preparations contained masked (or inactivated) binding sites that were activated by reaction with a ligand, or by exposure to high ionic strength; a behavior which is not uncommon for multisubunit proteins [30]. Both lectins contained lipid and carbohydrate (about 20% of each) and were extremely hydrophobic with a tendency to self-agglutinate, especially in the presence of calcium. They were high molecular weight aggregates, demonstrated by chromatography on Biogel P-300 and Sephacryl S-300 of purified lectins and crude hemolymph both in EDTA or in the presence of calcium. When

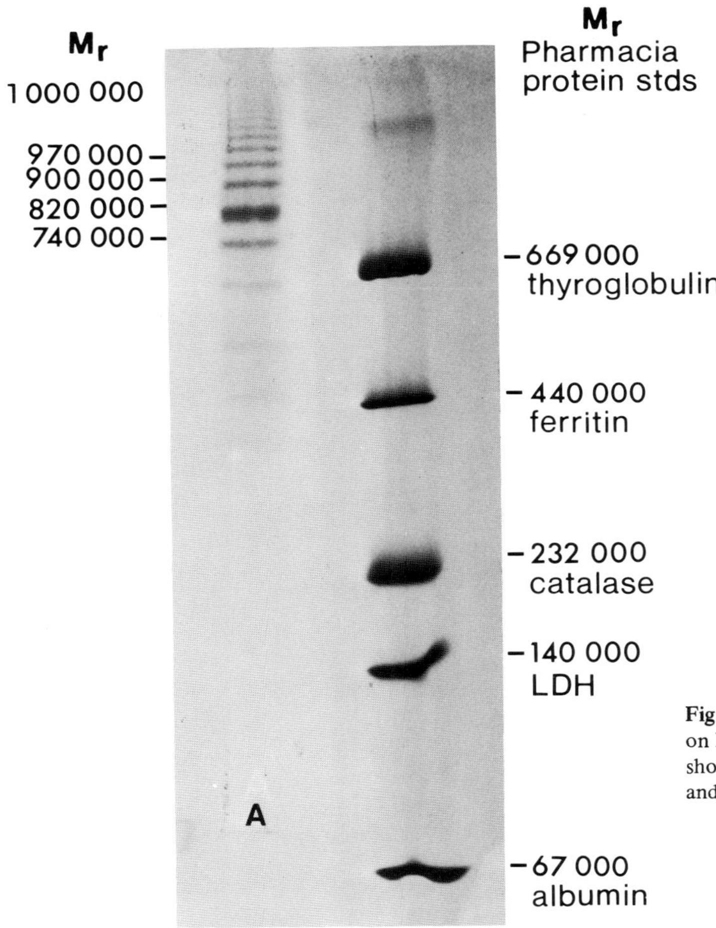

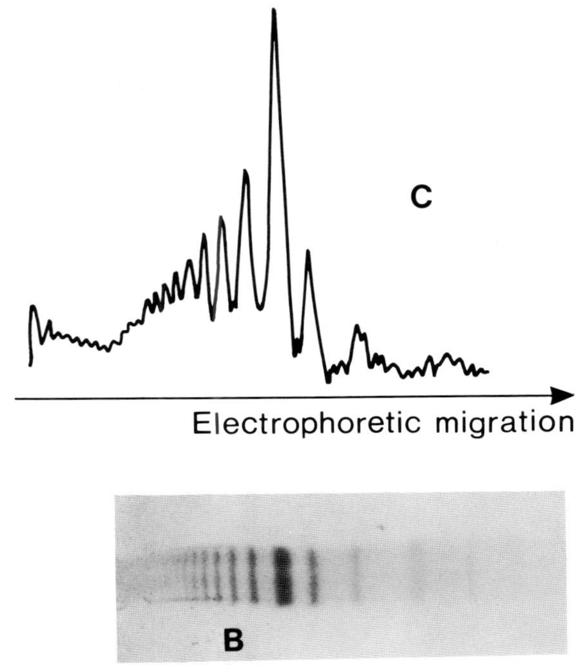

Fig. 1 A-C. Gradient-gel electrophoresis of purified Gigalin H + Gigalin E on Pharmacia gradient gels PAA 4/30 in Tris/borate/EDTA. Results show high M_r aggregates (**A** and **B**), Pharmacia HMW protein standards and (**C**) LKB laser densitometry of **B**

examined by gradient-gel electrophoresis in EDTA, the oyster lectins appeared as macromolecular aggregates in a series of distinct aggregation states with M_r ranging from about 500,000 to 1,600,000 (Fig. 1). These aggregates were unaffected by the addition of ligand, other proteins, or hemolymph, indicating that they were native (Olafsen J.A., Grant P.T., in preparation). The aggregates dissociated in 8 M urea. The resulting subunits could not agglutinate cells, but inhibited agglutination of human, though not horse erythrocytes. The subunits reaggregated easily, but never to form active complexes (Fig. 1).

SDS-PAGE (reducing conditions) demonstrated multiple subunits, i.e., α_1 (M_r 21,000), α_2 (M_r 22,500), and β (M_r 33,000, which could be split into two subunits of 15,500 and 17,500, respectively). Identification of subunits was carried out by use of immobilized horse or human erythrocyte ghosts, demonstrating that Gigalin H activity was predominated by $\alpha_1\alpha_2$, whereas Gigalin E activity was $\alpha_1\beta$. Although calcium did not appear necessary for binding to BSM, Gigalin E, but not Gigalin H, demonstrated a calcium requirement for the agglutination of erythrocytes.

A cell membrane associated lectin from the oyster *C. virginia* has been demonstrated [96] which seemed identical to a soluble lectin with high affinity for BSM, and probably for terminal NANA. The hemocyte membrane-associated lectin was considered to function as a membrane receptor in non-self recognition by molluscan hemocytes. Similarly Renwrantz and Stahmer [67] described a lectin in *Mytilus edulis* hemolymph which was inhibited by mucin. Not only the purified lectin, but also the addition of calcium ions significantly increased phagocytosis of yeast cells by *Mytilus* hemocytes. Location of recognition factors on the hemocyte surface was demonstrated and consequently both humoral and cell-bound lectin molecules were involved in the attachment of yeast cells to *Mytilus* hemocytes.

A β-galactosyl-binding lectin from *Tridacna maxima* [3] was a homogeneous glycoprotein (M_r 470,000) consisting of multiple subunits and requiring calcium for agglutination. It was suggested that the lectins in Tridacnidae served to digest their symbiont algae and photosynthetic products [54], and thus it is arguable that recognition of non-self might primarily be a nutrient-gaining process.

A lectin from the bivalve *Saxidomus purpuratus* purified by affinity chromatography agglutinated human (A) erythrocytes and was specific for N-acetyl-glucosamine and derivates. The native M_r was 40,000 consisting of M_r 14,000 subunits [83]. The lectin showed microheterogeneity (seven bands were demonstrated in PAGE), which was not attributable to carbohydrate groups, but probably caused by differences in amino acid composition. Purification by DEAE-cellulose resulted in only two active peaks (each M_r 40,000) with a single protein band from PAGE.

2.3 Crustacea

A lectin that agglutinated rabbit erythrocytes was purified from the American lobster *Homarus americanus* [18] and hemocytes were suggested as the source of this activity which unexpectedly increased when calcium was removed. The lectin was obtained from hemolymph and hemocyte lysate [36], resulting in two peaks of activity, one of which reacted with N-acetyl-D-galactosamine residues (L Ag-2), and the other (L Ag-1) reacted with NANA residues. Heterogeneity of binding sites was demonstrated [37].

McKay and Jenkin [52] demonstrated that uptake of erythrocytes by phagocytic cells from the crayfish *Cherax destructor* was dependent on hemolymph factors. In vivo elimination of bacteria from the hemolymph was promoted by opsonins which were depleted during this process, leading to a much slower secondary clearance [87]. Immunoadsorption revealed that the recognition factors were related, required calcium for activity, and were molecules of M_r 81,000 which could be dissociated into M_r 13,500 subunits (in [20]). Crayfish phagocytes also ingested bacteria in the absence of hemolymph [87].

2.4 Insecta

Insect defense mechanisms have recently been reviewed [49,64]. Many insect species contain agglutinins, and there are conflicting reports that activity may, or may not respond to antigenic challenge. There is, however, very little evidence that these agglutinins enhance recognition or phagocytosis by insect hemocytes. Opsonization only occurred when using a natural insect pathogen [64].

The beetle, *Allomyrina dichotoma* [89], contained two lectins; Allo A-I (M_r 65,000) consisting of two subunits (M_r 17,500 and 20,000, respectively) and Allo A-II (M_r 65,500), also with two (M_r 19,000 and 20,000) subunits. They were considered isolectins, and their biological function is not known, but increased activity was correlated with molt, injury, or age. Komano and associates [46], working with an inducible lectin in the flesh fly *Sarcophaga peregrina* larvae that agglutinated sheep erythrocytes and was inhibited by galactose and lactose, demonstrated the native M_r to be 190,000 consisting of four α (M_r 32,000) and two β (M_r 30,000) subunits. The lectin was apparently induced in two different physiological ways; by injury to the larval body wall and during pupation. It was demonstrated that hemolymph from normal larvae contained only the α subunit [47]. Since the tryptic maps of α and β subunits were very similar, the authors suggested, but did not actually show, that by injury to the body wall part of the α subunits were converted to β by a protease, resulting in the formation of an active $\alpha_4\beta_2$ lectin. There was no direct evidence that this lectin stimulated phagocytosis in *S. peregrina*.

2.5 Merostoma

The horseshoe crab *Limulus polyphemus* contains one of the best described lectin systems. The lectin Limulin, first purified and characterized by Marchalonis and Edelman [50] agglutinates a variety of erythrocytes (originally horse erythrocytes). It was specific for NANA, N-acetyl-glucosamine, and also bound KDO, and calcium was needed for agglutination. Limulin was a M_r 400,000 glycoprotein consisting of six noncovalently bound (M_r 67,000) subunits that could be further dissociated into M_r 22,500 subunits by 8 M urea; i.e., six basic units each composed of three subunits. Electron micrographs of the purified agglutinin demonstrated a uniform hexagonal ring structure [22]. Limulin was also extensively characterized by Roche and Monsigny [71] who reported a slightly different M_r (335,000 with M_r 19,000 subunits). The structure, function, and carbohydrate specificity of Limulin have been comprehensively treated [15]. Some interesting facts should, however, be noted. Limulin showed a marked tendency to self-aggregate through interaction of the carboxy-terminal segment of the subunits [45]. This aggregation is also noted after freeze-drying or concentration. Robey and Liu [69] claimed that Limulin was a C-reactive protein, with M_r 500,000 consisting of two nonidentical subunits (M_r 18,000 and 24,000). The two subunits were nonseparable, and amino acid composition suggested that the smaller subunit could have been derived from the other. Pistole and associates (see [62]) demonstrated that the serum from *L. polyphemus* also contained agglutinatting activity against a variety of bacteria. An LPS binding lectins with specifictiy for KDO [72] was claimed to be identical to Limulin. Later, a teichoic acid-binding lectin (agglutinating *Staphylococcus aureus*) was described that was distinct from Limulin [8]. Evidence was presented that this lectin was associated with hemocyanin (see Discussion).

Studies of the Japanese horseshoe crab *Tachypleus tridentatus* revealed the existence of heterogeneous lectins [79,80] that were purified by affinity chromatography on BSM-Sepharose 4B, and showed multiple species by SDS-PAGE (M_r 45,000; 42,000; 41,000; 39,000; 33,000; 29,000; 27,000 and 22,000). The affinity isolated agglutinins could be fractionated into four groups by gel chromatography, whereas whole hemolymph resulted in a single symmetrical peak corresponding to M_r 250,000, suggesting that the hemolymph agglutinin was eluted in a highly aggregated form. There was speculation that this might be due to the removal of an inhibitor during affinity gel isolation, but it could also be the result of intramolecular rearrangement (see Discussion). Thus, it is evident that binding of sialoconjugates is ubiquitous among the horseshoe crabs, and since these lectins usually also bind KDO, it has been suggested that they take part in defense against bacteria.

2.6 Arachnida

Multiple lectins were detected in the scorpion *Vaejovis confuscius* [95] and *Vaejovis spinigerus* [94] and also in the "whip scorpion" *Mastigoproctus giganteus* [92]. The lectins were apparently specific for sialic acids, their derivates, and sialoconjugates. The scorpion *Hadrurus arizonensis* also contained lectins that agglutinated a panel of erythrocytes and were heterogeneous in specificity, but predominantly bound sialoconjugates [93].

2.7 Chordata

The tunicates have been surveyed for lectin activity [20,97]. In most cases, multiple lectins have been detected, but few of these are well characterized, with some exceptions. *Halocynthia pyriformis* hemolymph contained a polymeric lectin (M_r 900,000) composed of noncovalently bound subunits of M_r 20,000. The binding specificity revealed the presence of two different binding sites within the same molecule; one specific for BSM, and the other specific for galactosyl-residues. Calcium was required for agglutination of horse, but not human erythrocytes [27,28].

Agglutinins for a variety of erythrocytes (with exceptionally high titer for guinea pig erythrocytes) were demonstrated in the ascidian *Botrylloides leachii* [74]. Two peaks of agglutinating activity were found following chromatography on Sephadex 200. One (HA-1) was specific for guinea pig erythrocytes in the presence of calcium, whereas the other (HA-2) agglutinated different types of erythrocytes without requiring divalent cations. Further characterization showed that the HA-1 agglutinin was a M_r 138,000 glycoprotein consisting of asymmetrically bound penta- or hexameric M_r 27,000 subunits (SDS-PAGE, reducing conditions). A lactose binding protein (LBP-3) was very similar to HA-1, both in molecular weight, antigenicity, and binding site specificity, but did not agglutinate guinea pig erythrocytes. The two proteins were copurified by affinity – or DEAE chromatography, but were separated on Sephacryl S-300 in EDTA. LBP-3 contained two covalently linked subunits S_1 and S_2 (M_r 27,000 and 22,000, respectively). The S_1 and HA-1 subunits appeared to be identical. HA-2 is a globular protein consisting of two noncovalently bound (probably identical) subunits (M_r 32–33,000). It was suggested that there is a receptor for the HA-2 molecule on the hemocyte, which mediates the attachment of sheep erythrocytes to hemocytes [20]. Attempts to enhance the agglutinin activity by injecting sheep or chicken erythrocytes resulted in a marginal, but significant rise in the titer of HA-2 after 2 weeks [16].

3 Discussion

Invertebrate lectins are undoubtedly involved in recognition events, and considerable biochemical information is now available on these lectins. Even if there are some common features; i.e., most invertebrate lectins are polymers of M_r 15–20,000 noncovalently bound subunits, no universal homology has hitherto been demonstrated. The situation is more one of diversity, which is not unexpected since the invertebrates are phylogenetically a so diverse group that they may have solved their problems of defense and nutrition in different ways.

Efforts to induce lectin activity in vivo in the hemolymph of invertebrates have only met with limited success. Following injection of bacteria or erythrocytes decreased agglutinin activity can usually be observed; interpreted as depletion of circulating recognition molecules following sudden introduction of a large dose of antigen. An exception was the annelid *Lumbricus terrestris* [17] that responded to the injection of erythrocytes by increasing the activity of circulating agglutinins. A more natural challenge than injection to a filter-feeder would be to introduce bacteria in the water [40]; this would provoke a response even if the triggering mechanism was outside the circulatory system, and no transient depletion would probably not result. Enhanced lectin activity in oyster hemolymph observed following in vivo exposure to bacteria [40] was indicative of a defense reaction. However, bacteria may be regarded as a potential food source for a filter-feeding bivalve, and as intracellular digestion plays a dominant role in these animals, one is led to speculate whether effective "eating" also accounts for effective defense. This is in part supported by the fact that in vivo exposure of oysters to algae (*Tetraselmis* sp. and *Phaeodactylum* sp.) also led to an increase, although more modest, of lectin activity in the hemolymph (Olafsen J.A., unpublished).

A striking observation was that bacteria pretreated (opsonized) with purified oyster lectin [40] were more rapidly engulfed by oyster hemocytes, whereas phagocytosis was not increased when bacteria and lectins were added simultaneously to the hemocyte suspension. Crude hemolymph, however, resulted in increased phagocytosis in both cases. A likely possibility was that conformational changes affecting the ratio of antigenic/cytophilic receptor sites rendered the affinity purified lectin more cytophilic than its hemolymph "precursor". The involvement of a "self-determinant" was also suggested by the fact that coating bacteria with purified *Limulus* lectin inhibited uptake in oyster hemocytes by 30%. However, considerable phagocytosis (up to about 50% of the experimental value) also took place in the absence of humoral factors (see Introduction). Indeed, it seems established that both humoral and cell-bound factors are involved, as clearly confirmed through the work of Renwrantz and colleagues [67] who demonstrated opsonizing activity of an isolated hemolymph lectin from *Mytilus*, as well as the involvement of hemocyte-bound lectins activated by calcium. This is also supported by the existence of identical lectins in hemolymph and on the plasma membrane of hemocytes [67,96].

A number of lectins need divalent cations (in particular calcium) for stabilization of the native molecule and also for the agglutination of cells. Multiple lectins in the hemolymph of an ascidian showed differential calcium requirements [74]. On the other hand, magnesium was shown to be inhibitory for the lectin from the cockroach [43]. Binding sites with differing calcium requirements within the same molecule have been reported for the tunicate lectin [27]; calcium was needed for agglutination of horse but not human erythrocytes and this also holds for the oyster lectins (Olafsen J.A., Fletcher T.C., Grant P.T., in prep.). Moreover, the oyster lectin, Gigalin E, was inhibited by zinc, and 100% inhibition was achieved at about the zinc concentration normally found in the hemolymph. However, inhibition was reversed by excess calcium, and the calcium/zinc ratio of the hemolymph would favor full Gigalin E activity. It has been demonstrated, however, that oyster hemocytes accumulate zinc [32, 83]. This makes it possible to speculate on an opsonin system influenced also by the intracellular/extracellular ratio of divalent cations.

Another functional approach would be the possible existence of an activating factor (i.e., protease). Many lectins have subunits existing in pairs of nearly similar M_r (i.e., differences of less than 2,000 daltons). Unfortunately, little data is available to indicate homology, but Komano and associates [47] inferred a protease that converted subunit a to β and activated the lectin by forming an active hexagonal $a_4\beta_2$ structure. Preliminary evidence from the *L. polyphemus* C-reactive protein also indicated transformation of the larger subunit to the smaller [69]. This possibility should be looked at in greater depth in other animals, too. Whereas it is generally agreed that lectins constitute only a small fraction of the total hemolymph protein, there are some conflicting reports. Acton [1] claimed, based mainly on the fact that absorption of oyster hemolymph with sheep erythrocytes removed most of the protein and all the agglutinating activity, that the *C. virginia* lectins accounted for 98% of the hemolymph protein. Referring to Jenkin's hypothesis on the spontaneous aggregation of functional recognition factors [44], and considering the diversity of antigenic determinants, one would expect that only a fraction of the biologically significant specificities could be picked up by an unnatural antigen such as the erythrocyte plasma membrane. In *C. gigas* we demonstrated by SDS-gradient-PAGE that the M_r 33,000 subunit coincides with a major hemolymph subunit pool, constituting 67% of the hemolymph protein. However, there is still no proof of true multispecificity of lectins in the hemolymph of any invertebrate.

The suggestion that an agglutinin to bacteria from *Limulus* was a hemocyanin [33] is intriguing. Arthropod hemocyanins are built from assemblies of one, two, four, or eight hexameric units; thus, *Limulus* hemocyanin is a 3.3 x 10^6 M_r protein containing 48 (M_r 70,000) subunits, whereas the lectin Limulin is a doubly stacked hexamer. The crayfish hemocyanin is composed of the random association into hexamers or dodecamers of three types of subunits, and the molecule occurs in a number of electrophoretically distinct forms stabilized by calcium (see [20]). Molluscan hemocyanins are assemblies of 10 or 20 subunits (M_r 400,000), and the aggregates appear as cylinders with a typical M_r of about 9 x 10^6 [6]. The stability of the aggregated hemocyanins is dependent on external conditions, such as pH, ionic strength, and divalent cations; changes may bring about dissociation into a series of distinct states of aggregation. The allosteric control of oxygen affinity in hemocyanins is apparently exerted through conformational changes of the protein.

In gastropods, agglutinins are not invariably present in all populations of a species; variations have also been found between different populations and even between individuals [60]. Finstad et al. [23] reported amino acid heterogeneity of Limulin, and speculated that this could introduce functional variability if the molecule was serving a defense purpose. Evidence for heterogeneity of oyster agglutinins was produced [51] by demonstrating that agglutination of sheep and rabbit erythrocytes by *C. virginica* hemolymph was inhibited by D-ribose (and was unaffected by calcium), whereas agglutination of human (A, B, and 0) erythrocytes was best inhibited by D-galactosamine (and was dependent on calcium ions). It was, thus, possible that oyster hemolymph contained a family of agglutinins each having slightly different specificity and reacting with structurally related sugars, and that they might be polymers each comprising a number of different subunits. Binding-site heterogeneity of lobster lectins was demonstrated by coagglutination of fluorescein conjugated human erythrocytes with mouse erythrocytes by L Ag-1 [37], whereas L Ag-2 agglutinated mouse erythrocytes only. Heterogeneity was also demonstrated by the fact that lectins from the Japanese horseshoe crab *T. tridentatus* showed activity maxima to human and horse erythrocytes in different peaks following chromatography according to molecular weight, and that their immunological properties were unrelated [80]. A number of reports are now available on both multiple lectins [37,92,94,95], isolectins [11,29,83,89,98], and also lectins with heterogeneous binding sites [20,27,28,37] in the hemolymph of different invertebrates. Some of these lectins are also present in highly aggregated forms [28, 53,60,80].

An intriguing observation is the biochemical similarity between lectins from the oyster *C. gigas* (Olafsen J.A., Fletcher T.C., Grant P.T., in prep.), the slug *L. flavus* [53], and the tunicate *H. pyriformis* [27,28]. Apart from the similar chemical specificity, the slug lectin behaved as a rapidly associating/dissociating polymeric system at high protein concentration, and in this it closely resembles the oyster lectins. The tunicate lectins are also closely related, both in calcium requirement, molecular size, and heterogeneity. Oyster lectins were demonstrated to be high molecular weight aggregates existing in a series of distinct aggregation states. The native molecule is constructed from three (or possibly four) kinds of subunits, expressing at least two functionally different activities. The association of subunits to biologically active complexes is probably not random since dissociated subunits readily reaggregate, but never to active molecules. Electron micrographs of intact or partly dissociated oyster lectins revealed ring-shaped (hexameric or pentameric) irregular structures (Olafsen J.A. and Pirie B.J.S., unpublished).

Although Ey and Jenkin [20] in 1982 pointed to the paucity of information related to the molecular basis of non-self recognition by invertebrate hemocytes, biochemical facts leading to a hypothetical model are emerging. It is, thus, tempting to hypothesize that at least some of these lectins would be stacked, ring-shaped (hexameric or pentameric) molecules with a hydrophobic core, possibly also containing lipid. Change in specificity could come about by rearrangement of subunits. In the hemolymph these lectins would exist as heterogenous, multisubunit proteins in a series of aggregation states, probably stabilized by hydrophobic forces, and separated into populations according to the predominantly expressed binding site. Since these high molecular weight lectins would operate through multipoint attachment, each bond could be weak, but agglutination might still come about through positive cooperativity, induced-fit, or allosteric changes. Thus, agglutination could be the result of an increase (above a threshold) of exposed binding sites. In fact, this could also explain the often reported failure of subunits to bind to ligands. Thus, cytophilic recognition following reaction to an antigen may possibly result from intramolecular rearrangement leading to an increase in cytophilic receptors.

Abbreviations

BSM	bovine submaxillary mucin
KDO	2-keto-3-deoxyoctanate
PAGE	polyacrylamide gel electrophoresis
SDS	sodium dodecyl sulphate
NANA	N-acetyl-neuraminic acid

Acknowledgments. Part of the experimental work referred to in this review was carried out during a stay at the NERC Institute of Marine Biochemistry, Aberdeen, Scotland. The discussion presented here would not have been possible without the help of three manuscripts in preparation with coauthors Prof. P.T. Grant and Dr. T.C. Fletcher from that Institute. The original information included in this article has partly resulted from research carried out with support from the Norwegian Fisheries Research Council. I would like to thank Mrs. Beryl Couzens for valuable help with the manuscript.

References

1. Acton RT, Bennett JC, Evans EE, Schrohenloher RE (1969) Physical and chemical characterization of an oyster hemagglutinin. J Biol Chem 244: 4128-4135
2. Anderson RS, Good RA (1976) Opsonic involvement in phagocytosis by molluscan hemocytes. J Invertebr Pathol 27: 57-64
3. Baldo BA, Sawyer WH, Stick RV, Uhlenbruck G (1978) Purification and characterization of a galactan-reactive agglutinin from the clam *Tridacna maxima* (Röding) and a study of its combining site. Biochem J 175: 467-477
4. Bang FB (1973) Immune reactions among marine and other invertebrates. Bioscience 23: 584-589
5. Barondes SH (1981) Lectins: their multiple endogenous cellular functions. Annu Rev Biochem 50: 207-231
6. Bonaventura C, Bonaventura J (1983) Respiratory pigments: structure and function. In: Hochachka PW (ed) Environmental biochemistry and physiology. Academic Press, New York, pp 1-50 (The mollusca, vol 2)
7. Boyd WC, Shapleigh E (1954) Antigenic relations of blood group antigen as suggested by tests with lectins. J Immunol 73: 226-231
8. Brandin ER, Pistole TG (1983) Polyphemin: a teichoic acidbinding lectin from the horseshoe crab, *Limulus polyphemus*. Biochem Biophys Res Commun 113: 611-617
9. Bretting H, Kabat EA (1976) Purification and characterisation of the agglutinins from the sponge *Axinella polypoides* and a study of their combining sites. Biochemistry 15: 3228-3236
10. Bretting H, Kabat EA, Liao J, Pereira MEA (1976) Purification and characterisation of the agglutinins from the sponge *Aaptos papillata* and a study of their combining sites. Biochemistry 15: 5029-5038
11. Bretting H, Stanislawski E, Jacobs G, Becker W (1983) Isolation and characterization of a lectin from the snail *Biomphalaria glabrata* and a study of its combining site. Biochem Biophys Acta 749: 143-152
12. Cheng TC (1981) Bivalves. In: Ratcliffe NA, Rowley AF (eds) Invertebrate blood cells, vol. 1 Academic Press, London, pp 233-300
13. Cheng TC (1984) Classification of molluscan hemocytes based on functional evidences. In: Cheng TC (ed) Invertebrate blood, cells and serum factors. Plenum, New York, pp 111-146 (Comparative pathobiology, vol 6)
14. Chorney MJ, Cheng TC (1980) Discrimination of self and non-self in invertebrates. In: Marchalonis JJ, Cohen N (eds) Self/non-self discrimination. Plenum, New York, pp 37-54 (Contemporary topics in immunobiology, vol 9)
15. Cohen E (1979) (ed) Biomedical applications of the horseshoe crab (Limulidae). Progr Clin Biol Res, vol 29. Liss, New York
16. Coombe DR, Ey PL, Jenkin CR (1982) Haemagglutinin levels in haemolymph from the colonial ascidian *Botrylloides leachii* following injection with sheep or chicken erythrocytes. Aust J Exp Biol Med Sci 60: 359-368
17. Cooper EL, Stein EA, Wojdani A (1984) Recognition receptors in Annelids. In: Cohen E (ed) Recognition proteins, receptors and probes: Invertebrates. Liss, New York, pp 43-54 (Prog Clin Biol Res, vol 157)
18. Cornick JW, Stewart JE (1973) Partial characterisation of a natural agglutinin in the haemolymph of the lobster (*Homarus americanus*). J Invertebr Pathol 21: 255-262
19. Dales RP (1978) Defence mechanisms. In: Mill PJ (ed) Physiology of Annelids. Academic Press, London, p 479
20. Ey PL, Jenkin CR (1982) Molecular basis of self/non-self discrimination in the invertebrata. In: Cohen H, Siegel MM (eds) The reticuloendothelial system. A comprehensive treatise. Plenum, New York London, pp 321-391 (Phylogeny and Ontogeny, vol 3)
21. Feng SY, Feng JS, Yamasu T (1977) Roles of *Mytilus coruscus* and *Crassostrea gigas* blood cells in defense and nutrition. Comp Pathobiol 3: 31-67
22. Fernandez-Moran H, Marchalonis JJ, Edelman GM (1968) Electron microscopy of a hemagglutinin from *Limulus polyphemus*. J Mol Biol 32: 467-469

23. Finstad CL, Good RA, Litman GW (1974) The erythrocyte agglutinin from *Limulus polyphemus* hemolymph: Molecular structure and biological function. Ann N Y Acad Sci 234: 170-182
24. Fitzgerald SW, Ratcliffe NA (1982) Evidence for the presence of subpopulations of *Arenicola marina* coelomocytes identified by their selective response towards Gram $^{+}$ve and Gram $^{-}$ve bacteria. Dev Comp Immunol 6: 23-34
25. Fletcher TC, Cooper-Willis CA (1982) Cellular defense systems of the mollusca. In: Cohen N, Sigel MM (eds) The reticuloendothelial system. A comprehensive treatise. Plenum, New York, pp 141-166 (Phylogeny and ontogeny, vol 3)
26. Foley DA, Cheng TC (1975) A quantitative study of phagocytosis by hemolymph cells of the pelecypods *Crassostrea virginica* and *Mercenaria mercenaria*. J Invertebr Pathol 25: 189-197
27. Form DM (1979) Isolation and characterization of a lectin from the hemolymph of a tunicate *Halocynthia pyriformis*. Thesis, Yale University, Massachusets
28. Form DM, Warr GW, Marchalonis JJ (1979) Isolation and characterization of a lectin from the hemolymph of a tunicate, *Halocynthia pyriformis*. Fed Proc 38: 934-941
29. Fountain DW, Campbell BA (1984) A lectin isolate from mucus of *Helix aspersa*. Comp Biochem Physiol 77B: 419-425
30. Friedman FK, Beychok S (1979) Probes of subunit assembly and reconstitution pathways in multisubunit proteins. Annu Rev Biochem 48: 217-250
31. Fries CR (1984) Protein hemolymph factors and their roles in invertebrate defense mechanisms: A review. In: Cheng TC (ed) Invertebrate blood, cells and serum factors. Plenum, New York, pp 49-109 (Comparative pathobiology, vol 6)
32. George SG, Pirie BJS, Cheyne AR, Coombs TL, Grant PT (1978) Detoxification of metals by marine bivalves: An ultrastructural study of the compartmentation of copper and zinc in the oyster *Ostrea edulis*. Marine Biol 45: 147-156
33. Gilbride KJ, Pistole TG (1981) The presence of copper in a purified lectin from *Limulus polyphemus:* Possible new role for hemocyanin. Dev Comp Immunol 5: 347-352
34. Gold ER, Balding P (1975) (eds) Receptor-specific proteins: Plant and animal lectins. Excerpta Medica, Amsterdam
35. Goldstein IJ, Hayes CE (1978) The lectins: Carbohydrate-binding proteins of plants and animals. Adv Carbohydr Chem Biochem 35: 127-340
36. Hall JL, Rowlands DT (1974a) Heterogeneity of lobster agglutinins I. Purification and physiochemical characterization. Biochemistry 13: 821-827
37. Hall JL, Rowlands DT (1974b) Heterogeneity of lobster agglutinins II. Specificity of agglutinin-erythrocyte binding. Biochemistry 13: 828-832
38. Hammarström S (1974) Structure, specificity, binding properties, and some biological activities of a blood group A-reactive hemagglutinin from the snail *Helix pomatia*. Ann NY Acad Sci 234: 183-197
39. Hammarström S, Westöö A, Björk I (1972) Subunit structure of *Helix pomatia* A hemagglutinin. Scand J Immunol 1: 295-309
40. Hardy SW, Fletcher TC, Olafsen JA (1977) Aspects of cellular and humoral defence mechanisms in the Pacific oyster, *Crassostrea gigas*. In: Solomon JB, Horton JD (eds) Developmental immunobiology. Elsevier/North Holland, Amsterdam, pp 59-66
41. Hardy SW, Grant PT, Fletcher TC (1977) A hemagglutinin in the tissue fluid of the Pacific oyster, *Crassostrea gigas*, with specificity for sialic acid residues in glycoproteins. Experientia (Basel) 33: 767-768
42. Harm H, Renwrantz L (1980) The inhibition of serum opsonins by a carbohydrate and the opsonizing effect of purified agglutinin on the clearance of nonself particles from the circulation of *Helix pomatia*. J Invertebr Pathol 36: 64-70
43. Ingram GA, East J, Molyneux DH (1983) Agglutinins of *Trypanosoma, Leishmania* and *Crithidia* in insect hemolymph. Dev Comp Immunol 7: 649-652
44. Jenkin CR, Hardy D (1975) Recognition factors of the crayfish and the generation of diversity. In: Hildeman WH, Benedict AA (eds) Immunologic phylogeny. Plenum, New York, pp 55-65
45. Kehoe JM, Kaplan R, Li SSL (1979) Functional implications of the covalent structure of Limulin: an overview. In: Cohen E (ed) Biomedical applications of the horseshoe crab (Limulidae). Liss, New York, pp 617-623

46. Komano H, Mizuno D, Natori S (1980) Purification of a lectin induced in the hemolymph of *Sarcophaga peregrina* larvae on injury. J Biol Chem 255: 2919-2924
47. Komano H, Mizuno D, Natori S (1981) A possible mechanism of induction of insect lectin. J Biol Chem 256: 7087-7089
48. Lackie AM (1980) Invertebrate immunity. Parasitology 80: 393-412
49. Lackie AM (1981) Immune recognition in insects. Dev Comp Immunol 5: 191-204
50. Marchalonis JJ, Edelman GM (1968) Isolation and characterisation of a haemagglutinin from *Limulus polyphemus*. J Mol Biol 32: 453-465
51. McDade JE, Tripp MR (1967) Mechanism of agglutination of red blood cells by oyster hemolymph. J Invertebr Pathol 9: 523-530
52. McKay D, Jenkin CR (1970) Immunity in the invertebrates: The role of serum factors in the phagocytosis of erythrocytes by haemocytes of the fresh water crayfish (*Parachaeraps bicarinatus*). Aust J Exp Biol Med Sci 48: 139-150
53. Miller RL, Collawn JR Jr, Fish WW (1982) Purification and macromolocular properties of a sialic acid-specific lectin from the slug *Limax flavus*. J Biol Chem 257: 7574-7580
54. Müller WEG, Zahn RK, Kurelec B, Lucu C, Müller I, Uhlenbruch G (1981) Lectin, a possible basis for symbiosis between bacteria and sponges. J Bacteriol 145: 548-558
55. Nicholson GL (1974) The interactions of lectins with animal cell surfaces. Int Rev Cytol 39: 89-190
56. Parish CR (1977) Simple model for self-non-self-discrimination in invertebrates. Nature 267: 711-713
57. Pauley GB (1974) Physicochemical properties of the natural agglutinins of some molluscs and crustaceans. Ann N Y Acad Sci 234: 145-160
58. Pauley GB, Granger GA, Krassner SM (1971) Characterization of a natural agglutinin present in the hemolymph of the California sea hare, *Aplysia californica*. J Invertebr Pathol 18: 207-218
59. Pauley GB, Krassner SM, Chapman FA (1971b) Bacterial clearance in the California sea hare, *Aplysia californica*. J Invertebr Pathol 18: 227-239
60. Pemberton RT (1974) Anti-A and anti-B of gastropods origin. Ann N Y Acad Sci 234: 95-121
61. Pereira MEA, Kabat EA (1979) Immunochemical studies on lectins and their application to the fractionation of blood group substances and cells. Crit Rev Immunol 1: 33-78
62. Pistole TG (1978) Broad-spectrum bacterial agglutinating activity in the serum of the horseshoe crab *Limulus polyphemus*. Dev Comp Immunol 2: 65-76
63. Prowse RH, Tait NN (1969) In vitro phagocytosis of amoebocytes from the hemolymph of *Helix aspersa* (Müller) I. Evidence for opsonic factor(s) in serum. Immunology 17: 437-443
64. Ratcliffe NA, Rowley AF (1984) Opsonic activity of insect hemolymph. In: Cheng TC (ed) Invertebrate blood, cells and serum factors. Plenum, New York, pp 187-204 (Comparative pathobiology, vol 6)
65. Renwrantz L (1983) Involvement of agglutinins (lectins) in invertebrate defense reactions: The immunobiological importance of carbohydrate-specific binding molecules. Dev Comp Immunol 7: 603-608
66. Renwrantz L, Mohr W (1978) Opsonizing effect of serum and albumin gland extract on the elimination of human erythrocytes from the circulation of *Helix pomatia*. J Invertebr Pathol 31: 164-170
67. Renwrantz L, Stahmer A (1983) Opsonizing properties of an isolated hemolymph agglutinin and demonstration of lectin-like recognition molecules at the surface of hemocytes from *Mytilus edulis*. J Comp Physiol 149: 535-546
68. Renwrantz L, Schäncke W,Harm H, Erl H, Liebsch H, Gercken J (1981) Discriminative ability and function of the immunobiological recognition system of the snail *Helix pomatia*. J Comp Physiol 141: 477-488
69. Robey FA, Liu T-Y (1981) Limulin: A C-reactive protein from *Limulus polyphemus*. J Biol Chem 256: 969-975
70. Robohm RA (1984) In vitro phagocytosis by molluscan hemocytes: A survey and critique of methods. In: Cheng TC (ed) Invertebrate blood, cells and serum factors. Plenum, New York, pp 147-172 (Comparative Pathobiology, vol 6)

71. Roche AC, Monsigny M (1974) Purification and properties of limulin: A lectin (agglutinin) from haemolymph of *Limulus polyphemus*. Biochim Biophys Acta 371: 242-254
72. Rostam-Abadi H, Pistole TG (1982) Lipopolysaccharide-binding lectin from the horseshoe crab, *Limulus polyphemus*, with specificity for 2-keto-3-deoxyoctonate (KDO). Dev Comp Immunol 6: 209-281
73. Rowley AF, Ratcliffe NA (1980) Insect erythrocyte agglutinins. In vitro opsonization experiments with *Clitumnus extradentatus* and *Periplaneta americana* hemocytes. Immunology 40: 483-492
74. Schluter SF, Ey PL, Keough DR, Jenkin CR (1981) Identification of two carbohydrate-specific erythrocyte agglutinins in the haemolymph of the protochordate *Botrylloides leachii*. Immunology 42: 241-250
75. Scott HT (1971) Recognition of foreignness in invertebrates II. In vitro studies of cockroach phagocytic hemocytes. Immunology 21: 817-828
76. Sharon N (1979) Possible functions of lectins in microorganisms, plants and animals. Glycocon Res, Proc 4th Int Symp Glycocon 1: 459-491
77. Sharon N (1984) Surface carbohydrates and surface lectins are recognition determinants in phagocytosis. Immunol Today (Amst) 5: 143-147
78. Sharon N, Lis H (1972) Lectins: Cell-agglutinating and sugarspecific proteins. Science 177: 949-959
79. Shimizu S, Ito M, Niwa M (1977) Lectins in the hemolymph of the Japanese horseshoe crab *Tachypleus tridentatus*. Biochim Biophys Acta 500: 71-79
80. Shishikura F, Sekiguchi K (1983) Agglutinins in the horseshoe crab hemolymph: Purification of a potent agglutinin of horse erythrocytes from the hemolymph of *Tachypleus tridentatus*, the Japanese horseshoe crab. J Biochem (Tokyo) 93: 1539-1546
81. Simkiss K, Mason AZ (1983) Metal ions: Metabolic and toxic effects. In: Hochachka PW (ed) The mollusca, vol 2. Environmental biochemistry and physiology. Academic Press, New York, pp 101-164
82. Sminia T, Winsemius AA, van der Knaap WPW (1981) Recognition of foreignness by blood cells of the freshwater snail *Lymnaea stagnalis*, with special reference to the role and structure of the cell coat. J Invertebr Pathol 38: 175-183
83. Tatsumi M, Arai Y, Itoh T (1982) Purification and characterization of a lectin from the shellfish *Saxidomus purpuratus*. J Biochem (Tokyo) 91: 1139-1146
84. Tripp MR (1966) Haemagglutinin in the blood of the oyster *Crassostrea virginica*. J Invertebr Pathol 8: 478-484
85. Tripp MR (1970) Defense mechanisms of molluscs. J Reticuloendothel Soc 7: 173-182
86. Tripp MR (1974) Oyster hemolymph proteins. Ann NY Acad Sci 234: 18-22
87. Tyson CJ, Jenkin DR (1974) Phagocytosis of bacteria in vitro by haemocytes from the crayfish (*Parachaeraps bicarinatus*). Aust J Exp Biol Med Sci 52: 341-348
88. Uhlenbruck G, Pardoe GI, Prokop O, Ishiyama I (1972) The serological specificity of snail agglutinins (protectins). Anim Blood Groups Biochem Genet 3: 125-139
89. Umetsu K, Kosaka S, Suzuki T (1984) Purification and characterization of a lectin from the beetle *Allomyrina dichotoma*. J Biochem (Tokyo) 95: 239-245
90. van der Knaap WPW, Boerrigter-Barendsen LH, van den Hoeven DSP, Sminia T (1981) Immunocytochemical demonstration of a humoral defense factor in blood cells (amoebocytes) of the pond snail, *Lymnaea stagnalis*. Cell Tissue Res 219: 291-296
91. van Oss CJ, Gillman CF (1972) Phagocytosis as a surface phenomenon II. Contact angles and phagocytosis of encapsulated bacteria before and after opsonization by specific antiserum and complement. J Reticuloendothel Soc 12: 497-502
92. Vasta GR, Cohen E (1984) Sialic acid-binding lectins in the "whip scorpion" *Mastigoproctus giganteus* serum. J Invertebr Pathol 43: 333-342
93. Vasta GR, Cohen E (1984) Characterization of the carbohydrate specificity of serum lectins from the scorpion *Hadrurus arizonensis* Stahnke. Comp Biochem Physiol 77B: 721-727
94. Vasta GR, Cohen E (1984) Serum lectins from the scorpion *Vaejovis spinigerus* Wood bind sialic acids. Experientia (Basel) 40: 485-487

95. Vasta GR, Cohen E (1984) Humoral lectins in the scorpion *Vaejovis confuscius:* A serological characterization. J Invertebr Pathol 43: 226-233
96. Vasta GR, Sullivan JT, Cheng TC, Marchalonis JJ, Warr GW (1982) A cell membrane-associated lectin of the oyster hemocyte. J Invertebr Pathol 40: 367-377
97. Vasta GR, Warr GW, Marchalonis JJ (1982) Tunicate lectins: Distribution and specificity. Comp Biochem Physiol 73: 887-900
98. Vretblad P, Hjorth R, Låås T (1979) The isolectins of *Helix pomatia*. Separation by isoelectric focusing and preliminary characterization. Biochim Biophys Acta 579: 52-61
99. Yeaton RW (1981) Invertebrate lectins I. Occurrence. Dev Comp Immunol 5: 391-402
100. Yeaton RW (1981) Invertebrate lectins II. Diversity of specificity, biological synthesis and function in recognition. Dev Comp Immunol 5: 535-545

Chapter 9 Immunorecognition in Invertebrates with Special Reference to Molluscs

T. Sminia[1] and W. P. W. van der Knaap[2]

CONTENTS

1 Introduction

All metazoan animals are protected against invading organisms by an internal defence system. This system comprises different elements. It consists of various cells, tissues and organs which enable organisms to react against potentially harmful pathogens from the external environment and against effete or derailed autologous material. Besides these structural components which form the cellular limb of the immune system diverse humoral factors produced by (several types of) blood cells are engaged in internal defence. In addition to lymphocytes, non-lymphoid cells, especially macrophage-like cells play a crucial role in the immune system (Fig. 1). Macrophages (a) are involved in the induction phase of the immune response (in the recognition of antigenic material and presentation of this material to lymphoid cells; (b) display a regulatory role in the immune response, and (c) play a part in the final phase of the immune reaction (uptake and degradation of antigenic material). The immune system can be divided in a non-specific and a specific part, the latter being characterized by memory formation. Depending on the phylogenetic development of the immune system only few, or more different cell types and humoral factors play a part in the immune response.

With respect to immunorecognition in invertebrates three basic questions can be asked: (1) which types of defence cells are present and what are their functions?;

1 Department of Histology,
2 Laboratory of Medical Parasitology, Faculty of Medicine, Free University, P.O. Box 7161, 1007 MC Amsterdam, The Netherlands

Immunity in Invertebrates
Ed. by M. Brehélin

(2) which types of factors occur and which role do they play? and (3) what are the interactions between cells and factors?

In this paper we will try to answer these questions by giving our ideas about the immune system of molluscs. The image that we have of their immune system is only partly based on the results of investigations on bivalves [8,52], and to a greater extent on studies of the gastropods *Biomphalaria glabrata* [4,7,54,55] and *Helix pomatia* [2, 31] and our own research of the freshwater snail *Lymnaea stagnalis* and will therefore reflect especially the situation in gastropods.

2 Defence Cells

In gastropods molluscs, several types of defence cells occur. These are antigen-trapping cells lining blood spaces [33], fixed phagocytes [30,42], resident or tissue haemocytes [38,46] and circulating haemocytes [2,38].

Antigen-trapping cells lining blood spaces [33] probably are endothelial cells, but their morphology has not been described. They play an important role in the first phase of the clearance of injected foreign particles. The cells are thought to bear carbohydrate receptors for haemolymph-coated foreign particles. After they have been trapped, the foreign particles are engulfed and then digested by circulating haemocytes.

Fixed phagocytes occur in the connective tissue and two types of truly fixed cells are present, viz. reticulum cells and pore cells (Fig. 2).

Reticulum cells possess numerous cytoplasmic processes that are closely associated with surrounding connective tissue fibrils. The cells have often large secondary lysosomes, a prominent Golgi apparatus and occasionally a dilated rough endoplasmic reticulum [42]. Reticulum cells should be regarded as true phagocytic cells as they ingest a wide range of injected foreign substances. Although no great differences seem to exist between reticulum cells and phagocytic circulating and resident haemocytes with respect to their phagocytic capacity, they may be functionally different. This supposition is based on the fact that haemocytes have the lysosomal enzyme peroxidase, whereas the reticulum cells lack this enzyme.

Pore cells possess a large round nucleus with conspicuous nucleoli, and contain numerous dark granules. The cell membrane of this cell type shows many vesicular and tubular invaginations, many of which are bridged by cytoplasmic tongues between which slits are present. In cross-sections of these areas the impression is gained that the cells have pores in their plasma membrane [38]. In these pores a three-dimensional sieve structure with holes of about 20–22 nm is present [6]. Pore cells are, in contrast to reticulum cells, selectively endocytotic. This capacity can be correlated with the structure of the cell membrane of the pore cell. It has been shown that the size of the holes of the sieve, present in the slits of the cell membrane, matches that of particles which can be ingested. Moreover, pore cells select on the basis of the nature of the material; the cells have a high affinity for proteinaceous material. It is conspicuous that in contrast to the avidity with which pore cells engulf foreign substances, the digestion rate of the incorporated material is low. This indicates that pore cells are not equipped with the appropriated enzymes, suggesting that the uptake of foreign material is not the prime function of these cells. Pore cells are presumably engaged in ingestion and turnover of self-proteins of the haemolymph.

Haemocytes of gastropods occur in the open circulatory system and can move into the loose connective tissue and vice versa. These amoeboid cells, which have a lysosomal system that may be developed to different extents, are actively involved in phagocytosis and encapsulation of foreign materials [26,28,38,41,46]. The resident or tissue haemocytes are usually less round and possess more and larger pseudopodia than those in the haemolymph, but they probably have similar functions. In the connective tissue of the heart-lung region, numerous resident haemocytes are present (Fig. 3). Depending on the snail species, specific areas of this connective tissue are regarded to constitute a haemopoietic organ [22,34]. The occurrence in these areas of stem cells or amoeboblasts has been reported and mitosis occurs frequently, especially when foreign material (parasites) is chronically present. Whether it is correct to name a tissue area where this occurs an Amoebocyte – Producing – Organ [15,18,22] remains a subject of debate, since haemocytes in circulation retain their capacity to divide [26, 39,47].

Circulating haemocytes have been studied much more intensively than the other cell types mentioned before, and much more is known of their role in internal defence. Therefore, the following section will deal with these cells in greater detail.

3 Circulating Haemocytes

Circulating haemocytes have been characterized by using morphological, enzyme- and immunocytochemical techniques and by using surface markers like lectins (see Chap. 2). In suspension, haemocytes have a spherical shape, in contact with (artificial) surfaces, the cells flatten and spread over the surface by extending cytoplasmic projections. A relative small number of cells maintain their round shape (Fig. 4). Intermediate forms between round and spreading cells are frequently observed. The round cells have the mophological characteristics of young cells, namely a high nucleus to cytoplasm ratio and numerous free (poly)ribosomes. The spreading cells have

Fig. 1. The main processes in the immunological defence mechanism of vertebrates

Fig. 2. Electron micrograph showing part of the cytoplasm of a reticulum cell (*A*) and a pore cell (*B*) of the freshwater snail *L. stagnalis. C* collagen fibrils; *GER* granular endoplasmic reticulum; *P* pores in the plasma membrane; *Pb* phagocytosed bacteria, x 20,000

Fig. 3. Haemocytes occur both free in the circulating haemolymph (*A*) and fixed in the connective tissue (*B*), (*L. stagnalis*), x 300

Fig. 4. Phase-contrast micrograph of haemocytes of *L. stagnalis* on glass. A round (*R*) and three spreading haemocytes. *G* cytoplasmic granules; *N* nucleus; *Ps* pseudopods, x 800

Fig. 5. Peroxidase activity in haemocytes of *L. stagnalis.* The number of peroxidase positive granules (*black granules*) varies greatly among the cells. *R* round haemocyte; *S* spreading haemocyte, x 800

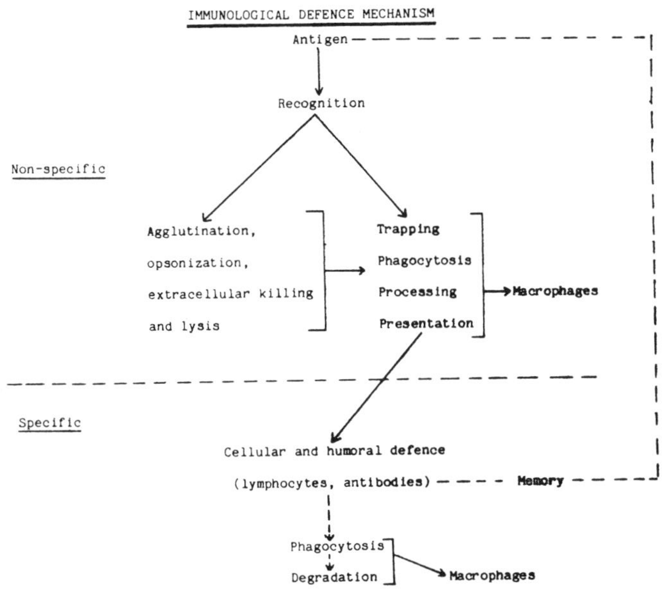
IMMUNOLOGICAL DEFENCE MECHANISM
Antigen
Recognition
Non-specific
Agglutination,
opsonization,
extracellular killing
and lysis
Trapping
Phagocytosis
Processing
Presentation
Macrophages
Specific
Cellular and humoral defence
(lymphocytes, antibodies)
Memory
Phagocytosis
Degradation
Macrophages

1

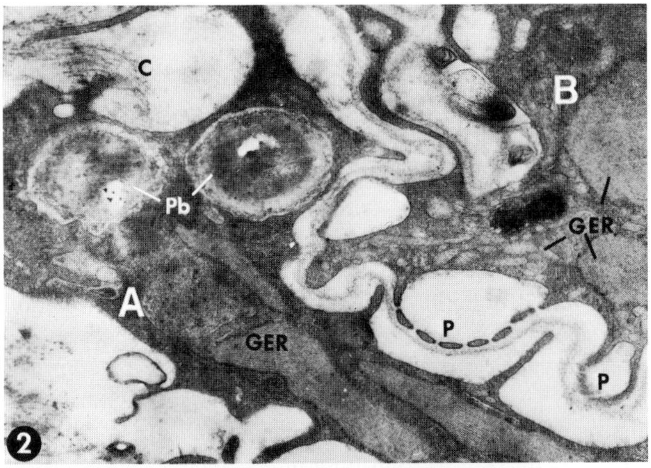
C
B
Pb
GER
A
GER
P
P

2

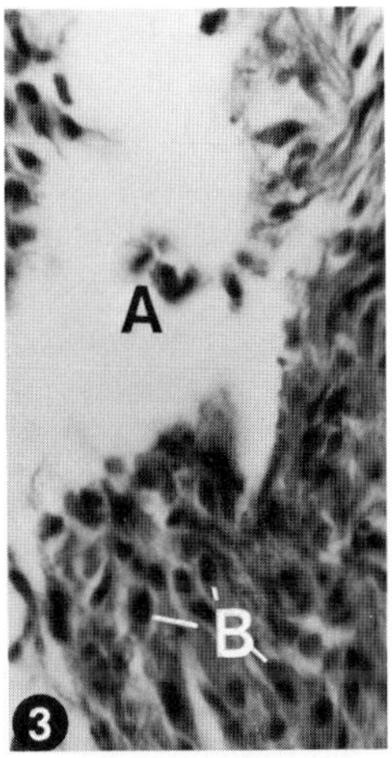
A
B

3

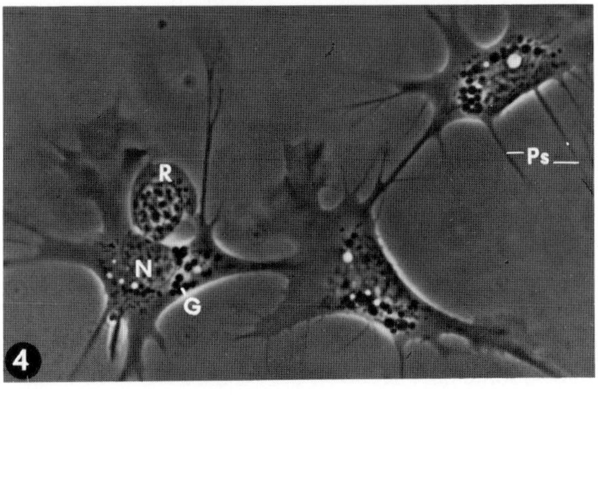
Ps
R
N
G

4

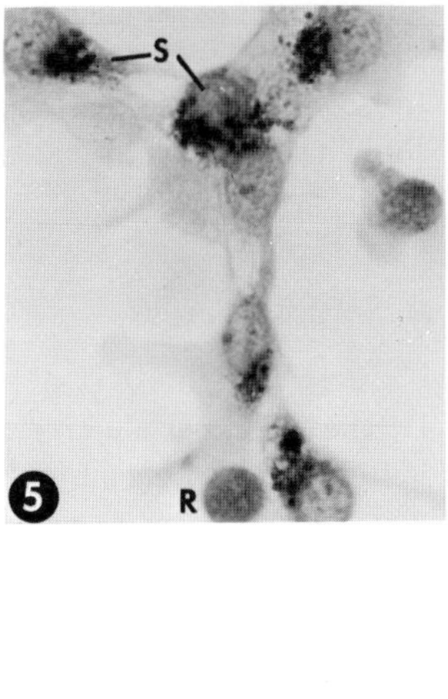
S
R

5

relatively more cytoplasm, an extensive Golgi apparatus and numerous lysosomes (granules). Thus, haemocytes are morphologically heterogeneous. As to the enzymes acid phosphatase, non-specific esterase and peroxidase the same heterogeneity exists (Fig. 5). The heterogeneity among haemocytes is also clear from the results of studies in which monoclonal antibodies, raised against snail haemocytes, were used for the staining of these cells ([56]; see also note added in proof). Some cells are positive with certain monoclonals while others are negative (Fig. 6).

Haemocytes are also functionally heterogeneous. To different extents, they perform tissue repair, they play an important role in clearing the haemolymph and connective tissue by phagocytosing foreign particles [38,47] or by encapsulating foreign materials [41] and they synthesize humoral defence factors [49]. The general conclusion from all these observations is that haemocytes constitute a morphologically and functionally heterogeneous population of cells. The question that arises is "are there different types of haemocytes, each with a specific function or is there only a single type of haemocyte, which carries out various functions?" The studies on haemocytes of *L. stagnalis* show that distinct morphologically different types of haemocytes, without intermediate stages, do not occur. Most likely, *L. stagnalis* has one basic type of haemocyte, which expresses various morphological forms (Fig. 7). The haemocytes pass through a differentiation series from round (young) cells to readily spreading (mature) cells. It can be assumed that the different developmental stages of haemocytes vary in functional capacities, but only quantitatively, not qualitatively. For example, spreading haemocytes show a higher phagocytic capacity than round haemocytes. Investigations on haemocyte morphology and surface membrane characteristics point to the possibility that several types of haemocytes would exist [17,26,56], that means that each cell type has its own function and progenitor cell. Up till now there is no consensus on the number of haemocyte types. As mentioned, haemocyte-producing organs in which stem cells are found have been described, but circulating haemocytes also have the capacity to divide. One of the ways to overcome the disagreement on haemocyte types, morphology and formation would be in our opinion the isolation of haemocytes and the characterization of the cells obtained. As the haemocytes differ in size and density, isolation can be carried out by conventional centrifugation and other separation techniques (see Chap. 1). But the recent development of monoclonal antibodies directed against snail haemocytes by Yoshino's group and also by ourselves will probably enable us to isolate the cells by using these monoclonal antibodies. After isolation and enrichment, the cells can be characterized morphologically and functionally. Moreover, it might even be possible by using monoclonals against haemocytes, to isolate precursors of haemocytes out of the haemocyte-producing organs and to study haemocyte proliferation and differentiation in vitro.

4 Serum Factors

Defence activities are exerted not only by defence cells. Substances that counteract invading foreign organisms or their products are present or may be secreted into the body fluid. Of these defence factors, some have received only little attention. Consequently, only little is known of their occurrence in different species, their synthesis site, molecular nature and mode of action; these factors are summarized here. In the haemolymph of *Helix aspersa*, a substance was demonstrated which inhibits the growth of nematodes [29]. A factor, probably of snail origin, which immobilizes miracidia of trematode parasites appears in the serum of *B. glabrata* after infection with such a parasite has taken place [23]. Lysins occur in the haemolymph, which are able to destroy foreign cells. Mammalian red blood cells are lysed by haemolymph of several planorbid species [25]. However, as *B. glabrata* is concerned, this is not a consistent phenomenon [7]; also of *L. stagnalis*, field-collected or laboratory-bred, only a low percentage of specimens displays this haemolytic activity (unpublished results). Haemolytic activity is present in the bivalve *Mytilus edulis*, and Wittke and Renwrantz [53] have shown that it is the haemocytes that secrete the lytic substance. It is conceivable that the membranes of cells, other than erythrocytes, will be sensitive to the lytic action. Possibly, it is the same mechanism that is responsible for the killing and degradation of schistosome parasites once these have been encapsulated by snail haemocytes [4]. Anti-bacterial factors occur; haemolymph of *L. stagnalis* has a selective bacteriostatic effect (unpublished results) and in three different abalone species, bactericidal activity could be induced by bacterial injections [14]. Also in abalones, anti-viral activity was detected in body fluids [20]. Serum of *H. pomatia* contains trypsin inhibitor [31] which may protect the snail against harmful proteolytic enzymes, secreted by infectious organisms.

Of the next group of defence factors, more is known, or to conceive, of the occurrence, nature and mode of action, viz. the lysosomal enzymes which are secreted by haemocytes as a response to foreign material. Lysozyme is found not only in bivalves [13] but also in snails [8,10,11,19]. Aminopeptidase [11,12] and lipase [9] have also been demonstrated in gastropod haemolymph.

The humoral defence activity to which most attention has been paid, is agglutination. Agglutinins, or lectins, cross-link foreign particles provided that these have the appropriate carbohydrate moieties on their surface for which the agglutinin has reactive sites. Interlinked micro-organisms are immobilized and thus an easy prey to defence cells. Although one should realize that the prime function of agglutinins might well be in metabolism (see [27]), it is tempting to assume that these molecules, which seem to occur in all gastropod species, play their major role in internal defence. From whatever point of view agglutinin is observed, heterogeneity is the key word. The molecules generally are glycoproteins composed of subunits, but the molecular weights and the numbers of subunits differ greatly between the different species. Also, the specificities for carbohydrates differ largely between the various agglutinins, the majority having broad specificities (e.g. *L. stagnalis,* [48]; *B. glabrata,* [7]). As a result, a wide variety of foreign cells is agglutinated (e.g. [48]), suggesting that a not very sensitive recognition system is at stake. However, foreign cells of one specific type, e.g. human A1 erythrocytes, are agglutinated by the serum of some specimens of one single stock of snails, and not by others [25,48]; also it occurs that snail haemolymph does agglutinate one batch of foreign cells of a specific type, and not another batch of the same type [16,47]. This indicates that the agglutinins constitute a very sensitive immunorecognition system.

Opsonization is the final humoral defence activity that we mention. Since Prowse and Tait [28] had shown that the serum of *H. aspersa* contains a substance which renders foreign cells more suitable for phagocytosis by haemocytes, many reports describing opsonin in gastropods have followed (see e.g. [43]). Both in vitro experiments (e.g. [50]) and in vivo studies (e.g. [33]) have shown that it is agglutinin molecules that bind to sugar moieties on the foreign cells and subsequently link these to the surface of defence cells. Then, cellular activities do the rest. The interplay between foreign object, serum factors and the surface of defence cells is the subject of the next section.

5 Interactions Between Haemocytes and Serum Factors

So, we have haemocytes which wall off and engulf foreign material, and serum factors with opsonizing and agglutinating activity. The question arises, how do they cooperate? Since in gastropods, as in other investigated invertebrates, no antibodies have been found, the main type of phagocytosis must be non-immunological. Recognition of foreignness in non-immunological phagocytosis is based on physico-chemical factors like wettability and surface charge of both phagocytic cells and foreign particles. This enables haemocytes to adhere to glass [43] and to engulf inert particles [38]. Moreover, lectins with specific binding sites for carbohydrate moieties are involved. These substances appear to play a crucial role in the recognition, adhesion, walling off and engulfment of foreign material [33,50].

Like other invertebrate and vertebrate cells, molluscan haemocytes have a sugar coating which consists of the oligosaccharide units of membrane glycoproteins and glycolipids. Certain plant lectins bind to the haemocytes, and this binding is inhibited or reversed by the sugars for which the lectins are specific. This has given information on which sugar receptors occur on the surface of haemocytes of *H. pomatia* [32], *B. glabrata* [54] and *L. stagnalis* [44]. Also by other methods like microhaemadsorption assays the occurrence of specific sugars has been demonstrated on the surface membrane of snail haemocytes [36].

In conclusion, these findings reveal sugar-containing, lectin-binding sites on molluscan haemocytes. Whether the native (snail) lectins bind to the haemocytes through this mechanism, is still unclear. However, as trypsin treatment of *L. stagnalis* haemocytes abolishes their capacity to bind both ConA [44] and native lectins [50], this mechanism seems to be quite plausible.

Fig. 6. Haemocytes of *L. stagnalis* stained with a two-step immunoperoxidase reaction, using a monoclonal antibody raised against circulating haemocytes. Both stained and unstained cells are present, x 800

Fig. 7. The ontogenetic stages of the haemocytes of gastropods

Fig. 8. Gastropod haemocyte in internal defence. *1* synthesis of lectin; *2* secretion of lectin; *3* physico-chemical attraction; *4* receptor: lectin as integral part of the cell-surface; *5* receptor: cytophilic lectin on unknown binding site; *6* receptor: cytophilic lectin on surface carbohydrate; *7* opsonization, then adhesion; *8* agglutination, then adhesion; *9* internalization; *10* secretion/leakage of lysosomal enzymes

Fig. 9. The main events in the internal defence system of molluscs (for comparison with the vertebrate immune system see Fig. 1)

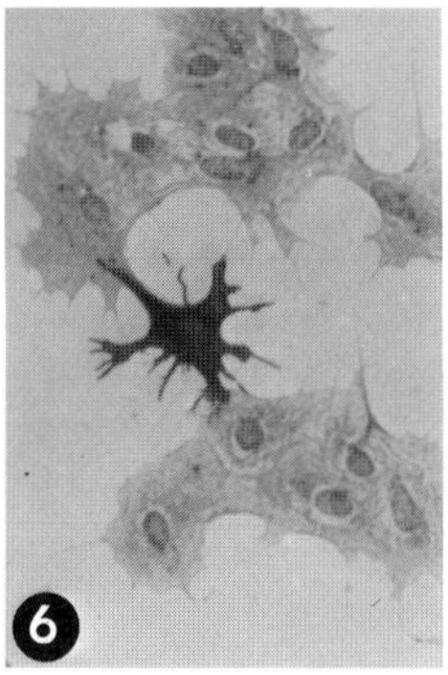

6

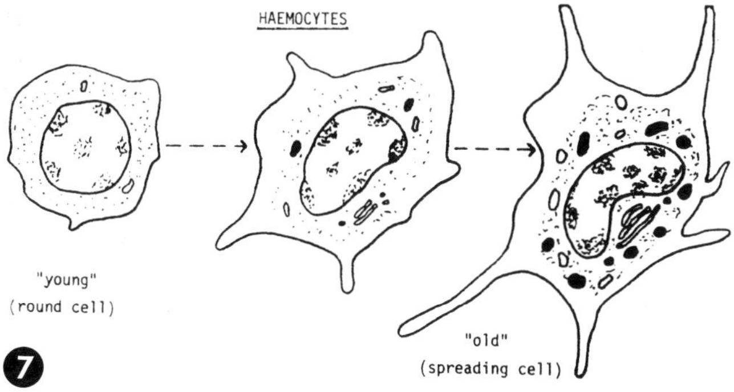
HAEMOCYTES
"young"
(round cell)
"old"
(spreading cell)

7

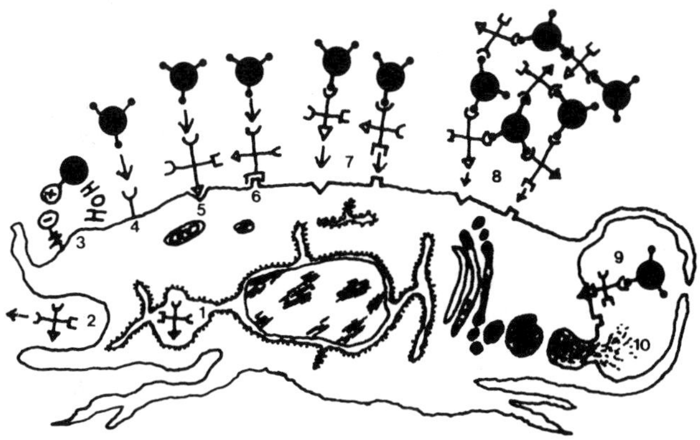
1
2
3
4
5
6
7
8
9
10

8

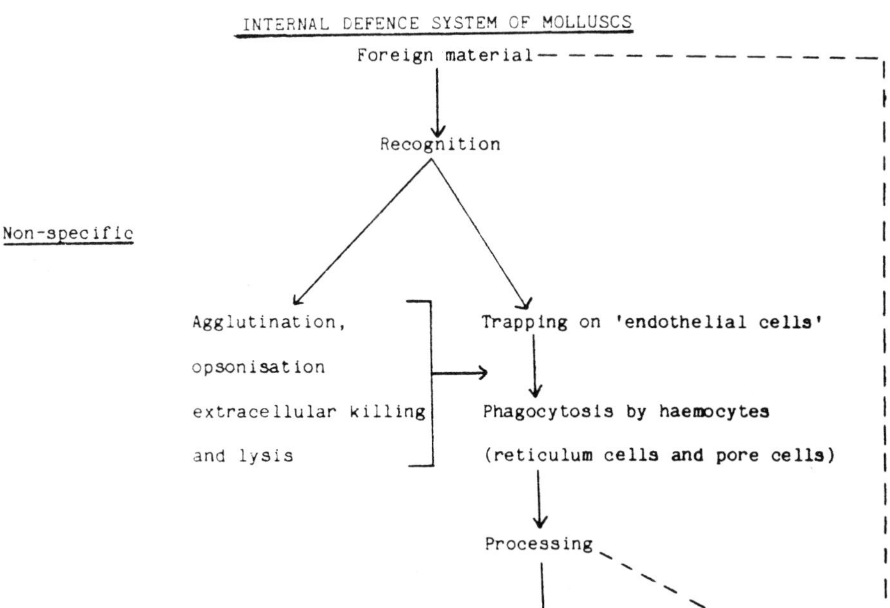
INTERNAL DEFENCE SYSTEM OF MOLLUSCS
Foreign material
Recognition
Non-specific
Agglutination,
opsonisation
extracellular killing
and lysis
Trapping on 'endothelial cells'
Phagocytosis by haemocytes
(reticulum cells and pore cells)
Processing
Degradation
'Memory'

9

The haemocytes not only have receptors for circulating lectin molecules, lectins appear to be constantly present on the surface of the cells. As for *L. stagnalis* haemocytes, this has been demonstrated immunocytochemically with the use of polyvalent anti-lectin antibodies [49]. Monoclonal antibodies to snail constituents, now becoming available ([56]; see also note added in proof) may help to study this phenomenon in more detail. The cell surface lectins may be an integral part of the cell envelope. Possibly in addition, surface receptors for foreignness can be detached from, and reattached to the cell surface, as was demonstrated in functional tests with haemocytes and native lectins [50].

There are indications that native lectins may play a role in the haemocyte's capacity to respond to chemotactic stimuli [35]. Moreover, it has been reported that snail haemocytes display membrane plasticity [55], become altered in behaviour and form funnel-like pseudopods [37] upon contact with non-native lectins. All these alterations can be envisaged to facilitate endocytosis of foreign material, and they may indicate how native lectins stimulate phagocytosis.

With respect to the interactions between gastropod haemocytes and opsonizing and agglutinating substances, we can propose the following model for the role of gastropod haemocytes in internal defence (Fig. 8). The haemocytes bind foreign objects through physico-chemical forces. Moreover they synthesize and secrete factors with opsonizing and agglutinating properties; recognition takes place via these factors, that are cell-bound and free in the serum. After recognition, adhesion and engulfment take place. Thus, the haemocyte has a very central position in the internal defence system.

Comparing the internal defence system of molluscs with the immunological defence system of vertebrates, one can conclude that the specific part of the defence system is lacking. The macrophage-like haemocytes which process the foreign material are the only haemolymph cells present, indicating that presentation of the foreign material to lymphocyte-like cells does not occur (Fig. 9).

6 Actively Acquired Immunity

In the foregoing we have described and discussed innate defence reactions in molluscs. In this section we will make some remarks on actively acquired heightened responsiveness in gastropod molluscs.

Several studies have been done in which possible alterations were measured of humoral and cellular components of the defence system, resulting from previous exposure to foreign material. Altered serum levels of several types of humoral defence factors have been demonstrated. Lysosomal enzymes, most probably of haemocyte origin, are found at increased levels in the haemolymph of several molluscan species [10, 11]. The induction of bactericidal substances has been reported [14] as were increased titers of agglutinin [16], opsonin and anti-parasitic substances [23]. Upon repeated injection or implantation of foreign material the phagocytic and encapsulation activity of the recipient increased [1,3,21,47]. Moreover, the proliferation [18,34,47] and differentiation of haemocytes (e.g. increase of the cell surface due to more and larger cytoplasmic projections, increase of the phagocytic capacity [47]) altered upon contact with foreign material. These results suggest that both the total number of defence cells and the activity of the individual cells increased after the first contact with foreign material.

Acquired immunity in vertebrates, a state of increased defence activity due to previous contact with foreign (antigenic) material, is characterized by specificity and memory formation. Regarding specificity we must realize that this phenomenon is based on the presence of highly specific anti-cells (T-lymphocytes) and antibodies, produced by B-lymphocyte derived plasma cells, which cooperate with a third population of cells, the macrophages. In molluscs, only this third element is present: the haemocytes are in certain aspects morphologically and functionally comparable to vertebrate macrophages. Since there are no specific anti-cells and no specific antibodies – only lectin-like serum factors are present, substances with a relatively broad spectrum specificity – one has to come to the conclusion that anamnestic responses in molluscs and presumably also in other invertebrate groups, must be based on low specificity. Indeed, the observed activation of the defence system is generally non-specific (see e.g. [24,46]).

As to memory in adaptive responses, the following can be said. In *L. stagnalis* the induced activation of the defence system is relatively long-lasting (at least 64 days; [47]), whereas in other molluscs such activation could still be measured about 1 month after induction [1,3]. As most probably these induced reactions are non-specific it is better not to use the term "immunological memory", but descriptions such as "duration of the induced activation".

7 Summary and Concluding Remarks

Gastropod molluscs possess several types of defence cells viz. fixed reticulum cells, pore cells and mobile haemocytes which occur both in the haemolymph and the connective tissue. The reticulum cells are avidly phagocytic, whereas the pore cells are selectively endocytic. The haemocytes display several functions as phagocytosis, encapsulation, wound repair and the synthesis of humoral defence factors. On morphological grounds several subtypes of haemocytes can be distinguished. In order to find out whether the morphological heterogeneity reflects functional differences among the haemocytes, it is necessary to isolate the haemocyte subpopulations and to test the fractions functionally.

Diverse types of serum factors occur such as agglutinins, opsonins, bactericidins, anti-parasitic factors and lytic enzymes. Cells and factors together, and working in cooperation, constitute an innate defence system that is able to efficiently deal with foreign substances, cells and organisms. A scheme for the central role of haemocytes in internal defence is presented.

Overlooking the data on adaptive defence reactions in molluscs, one must come to the conclusion that a non-specific rather than a specific elevation of the defence system occurs upon contact with foreign material, and that confirmation of the occurrence of immunological memory in the classical sense is still lacking.

References

1. Acton RT, Evans EE (1968) Bacteriophage clearance in the oyster (*Crassostrea virginica*). J Bacteriol 95: 1260-1266
2. Bayne CJ (1973) Molluscan internal defense mechanism: the fate of C^{14}-labeled bacteria in the land snail *Helix pomatia* (L). J Comp Physiol 86: 17-25
3. Bayne CJ (1980) Molluscan immunity: induction of elevated immunity in the land snail (*Helix)* by injections of bacteria (*Pseudomonas aeruginosa*). Dev Comp Immunol 4: 43-54
4. Bayne CJ, Buckley PM, Dewan PC (1980) Macrophage-like hemocytes of resistant *Biomphalaria glabrata* are cytotoxic for sporocysts of *Schistosoma mansoni* in vitro. J Parasitol 39: 413-419
5. Bayne CJ, Sminia T, Van der Knaap WPW (1980) Immunological memory: status of molluscan studies. In: Manning MJ (ed) Phylogeny of immunological memory, Elsevier/North Holland, Biomedical Press, Amsterdam, pp 57-64
6. Boer HH, Sminia T (1976) Sieve structures of slit diaphragms of podocytes and pore cells of gastropod molluscs. Cell Tissue Res 170: 221-229
7. Boswell CA, Bayne CJ (1984) Isolation, characterization and functional assessment of a hemagglutinin from the plasma of *Biomphalaria glabrata*, intermediate host of *Schistosoma mansoni*. Dev Comp Immunol 8: 559-568
8. Cheng TC (1978) The role of lysosomal hydrolases in molluscan cellular response to immunological challenge. Comp Pathobiol 4: 59-71
9. Cheng TC, Yoshino TP (1976) Lipase activity in the hemolymph of *Biomphalaria glabrata* (Mollusca) challenged with bacterial lipids. J Invertebr Pathol 28: 143-146
10. Cheng TC, Chorney MJ, Yoshino TP (1977) Lysozyme-like activity in the hemolymph of *Biomphalaria glabrata* challenged with bacteria. J Invertebr Pathol 29: 170-174
11. Cheng TC, Guida VG, Gerhart PL (1978) Aminopeptidase and lysozyme activity levels and serum protein concentrations in *Biomphalaria glabrata* (Mollusca) challenged with bacteria. J Invertebr Pathol 32: 297-302
12. Cheng TC, Howland KH, Moran HJ, Sullivan JT (1983) Studies on parasitic castration: aminopeptidase activity levels and protein concentrations in *Illyanassa obsoleta* (Mollusca) parasitized by larval trematodes. J Invertebr Pathol 42: 42-50
13. Cheng TC, Rodrick GE, Foley DA, Koehler SA (1975) Release of lysozyme from hemolymph cells of *Mercenaria mercenaria* during phagocytosis. J Invertebr Pathol 25: 261-265
14. Cushing JE, Evans EE, Evans ML (1971) Induced bactericidal responses of abalones. J Invertebr Pathol 17: 446-448
15. Jeong KH, Lie KJ, Heyneman D (1983) The ultrastructure of the amebocyte-producing organ in *Biomphalaria glabrata*. Dev Comp Immunol 7: 217-228
16. Jeong KH, Sussman S, Rosen SD, Lie KJ, Heyneman D (1981) Distribution and variation of hemagglutinating activity in the haemolymph of *Biomphalaria glabrata*. J Invertebr Pathol 38: 256-263
17. Joky A, Matricon-Gondran M, Benex J (1983) Fine structural differences in the amoebocytes of *Biomphalaria glabrata*. Dev Comp Immunol 7: 669-672
18. Joky A, Matricon-Gondran M, Benex J (1985) Response to the amoebocyte-producing organ of sensitized *Biomphalaria glabrata* after exposure to *Echinostoma caproni* miracidia. J Invertebr Pathol 45: 28-33
19. Kassim OO, Richards CS (1978) *Schistosoma mansoni*: lysozyme activity in *Biomphalaria glabrata* during infection with two strains. Exp Parasitol 46: 213-217
20. Li CP, Prescott B, Johnes WG (1962) Antiviral activity of a fraction of abalone juice. Proc Soc Exp Biol Med 109: 534-538
21. Lie KJ, Heyneman D, Lim HK (1975) Studies on resistance in snails: specific resistance induced by irradiated miracidia of *Echinostoma lindoense* in *Biomphalaria glabrata* snails. J Parasitol 5: 627-631
22. Lie KJ, Heyneman D, Yau P (1975) The origin of amebocytes in *Biomphalaria glabrata*. J Parasitol 61: 574-576

23. Lie KJ, Jeong KH, Heyneman D (1980) Inducement of miracidia-immobilizing substance in the haemolymph of *Biomphalaria glabrata*. Int J Parasitol 10: 183-188
24. Lie KJ, Jeong KH, Heyneman D (1982) Further characterization of acquired resistance in *Biomphalaria glabrata*. J Parasitol 68: 529-531
25. Michelson EH, Dubois L (1977) Agglutinins and lysins in the molluscan family Planorbidae: a survey of hemolymph, eggmasses, and albumen-gland extracts. Biol Bull (Woods Hole) 153: 219-227
26. Ottaviani E (1983) The blood cells of the freshwater snail *Planorbis corneus* (Gastropoda, Pulmonata). Dev Comp Immunol 7: 209-216
27. Parish CR (1977) Simple model for self-non-self-discrimination in invertebrates. Nature 267: 711-713
28. Prowse RH, Tait NN (1969) In vitro phagocytosis by amoebocytes from the haemolymph of *Helix aspersa* (Müller) I. Evidence for opsonic factor(s) in serum. Immunology 17: 437-443
29. Ratanarat-Brockelman C (1977) Isolation of nematode inhibitor from hemolymph of the snail *Helix aspersa*. Biol Bull (Woods Hole) 152: 406-414
30. Reade PC (1968) Phagocytosis in invertebrates. Aust J Exp Biol Med Sci 46: 219-229
31. Renwrantz L (1979) Eine Untersuchung molekularer und zellulärer Bestandteile der Hämolymphe von *Helix pomatia* unter besonderer Berücksichtigung immunobiologisch aktiver Komponenten. Zool Jahrb Physiol 83: 283-333
32. Renwrantz LR, Cheng TC (1977) Identification of agglutinin receptors on haemocytes of *Helix pomatia*. J Invertebr Pathol 29: 88-96
33. Renwrantz LR, Schäncke W, Harm H, Erl H, Liebsch H, Gercken J (1981) Discriminative ability and function of the immunobiological recognition system of the snail *Helix pomatia*. J Comp Physiol 141: 447-488
34. Rondelaud D, Barthe D (1981) The development of the amoebocyte-producing organ in *Lymnaea truncatula* (Müller) by *Fasciola hepatica* (L). Z Parasitenkd 65: 331-341
35. Schmid LS (1975) Chemotaxis of hemocytes from the snail *Viviparus malleatus*. J Invertebr Pathol 25: 125-131
36. Schoenberg DA, Cheng TC (1980) Lectin-binding specificities of hemocytes from two strains of *Biomphalaria glabrata* as determined by microhemadsorption assays. Dev Comp Immunol 4: 617-628
37. Schoenberg DA, Cheng TC (1981) The behaviour of *Biomphalaria glabrata* (Gastropoda: Pulmonata) hemocytes following exposure to lectins. Trans Am Microsc Soc 100: 345-358
38. Sminia T (1972) Structure and function of blood and connective tissue cells of the freshwater pulmonate *Lymnaea stagnalis* studied by electron microscopy and enzyme histochemistry. Z Zellforsch Mikrosk Anat 130: 497-526
39. Sminia T (1974) Haemopoiesis in the freshwater snail *Lymnaea stagnalis* studied by electron microscopy and autoradiography. Cell Tissue Res 150: 443-454
40. Sminia T, Barendsen L (1980) A comparative morphological and enzyme histochemical study on blood cells of the freshwater snails *Lymnaea stagnalis*, *Biomphalaria glabrata* and *Bulinus truncatus*. J Morphol 165: 31-37
41. Sminia T, Borghart-Reinders E, Van de Linde AW (1974) Encapsulation of foreign materials experimentally introduced into the freshwater snail *Lymnaea stagnalis*. An electronmicroscopic and autoradiographic study. Cell Tissue Res 153: 307-326
42. Sminia T, Van der Knaap WPW, Kroese FGM (1979) Fixed phagocytes in the freshwater snail *Lymnaea stagnalis*. Cell Tissue Res 196: 545-548
43. Sminia T, Van der Knaap WPW, Edelenbosch P (1979) The role of serum factors in phagocytosis of foreign particles by blood cells of the freshwater snail *Lymnaea stagnalis*. Dev Comp Immunol 3: 37-44
44. Sminia T, Winsemius AA, Van der Knaap WPW (1981) Recognition of foreignness by blood cells of the freshwater snail *Lymnaea stagnalis*, with special reference to the role and structure of the cell coat. J Invertebr Pathol 38: 175-183
45. Stanislawsky E, Renwrantz L, Becker W (1976) Soluble blood group reactive substances in the hemolymph of *Biomphalaria glabrata* (Mollusca). J Invertebr Pathol 28: 301-308
46. Tripp MR (1970) Defense mechanisms of mollusks. J Reticuloendothel Soc 7: 173-182

47. Van der Knaap WPW, Boots AMH, Van Asselt LA, Sminia T (1983) Specificity and memory in increased defence reactions against bacteria in the pond snail *Lymnaea stagnalis*. Dev Comp Immunol 7: 435-443
48. Van der Knaap WPW, Doderer A, Boerrigter-Barendsen LH, Sminia T (1982) Some properties of an agglutinin in the haemolymph of the pond snail *Lymnaea stagnalis*. Biol Bull (Woods Hole) 162: 404-412
49. Van der Knaap WPW, Boerrigter-Barendsen LH, Van den Hoeven DSP, Sminia T (1981) Immunocytochemical demonstration of a humoral defence factor in blood cells (amoebocytes) of the pond snail *Lymnaea stagnalis*. Cell Tissue Res 219: 292-296
50. Van der Knaap WPW, Sminia T, Schutte R, Boerrigter-Barendsen LH (1983) Cytophilic receptors for foreignness and some factors which influence phagocytosis by invertebrate leucocytes: in vitro phagocytosis by amoebocytes of the pond snail *Lymnaea stagnalis*. Immunology 48: 377-383
51. Van der Knaap WPW, Tensen CP, Kroese FGM, Boerrigter-Barendsen LH (1982) Adaptive defence reactions against bacteria in the pond snail *Lymnaea stagnalis*. Dev Comp Immunol 6: 775-780
52. Vasta GR, Sullivan JT, Cheng TC, Marchalonis JJ, Warr GW (1982) A cell membrane-associated lectin of the oyster hemocyte. J Invertebr Pathol 40: 367-377
53. Wittke M, Renwrantz L (1984) Quantification of cytotoxic hemocytes of *Mytilus edulis* using a cytotoxicity assay in agar. J Invertebr Pathol 43: 248-253
54. Yoshino TP (1981) Concanavalin A-induced receptor redistribution on *Biomphalaria glabrata* hemocytes: characterization of capping and patching responses. J Invertebr Pathol 38: 102-112
55. Yoshino TP (1982) Lectin-induced modulation of snail hemocyte surface determinants: clearance of ConA-receptor complexes. Dev Comp Immunol 6: 451-461
56. Yoshino TP, Granath Jr WO (1983) Identification of antigenically distinct hemocyte subpopulations in *Biomphalaria glabrata* (Gastropoda) using monoclonal antibodies to surface membrane markers. Cell Tissue Res 232: 553-564

Note Added in Proof (appeared after completion of the manuscript):

Dikkeboom R, Van der Knaap WPW, Maaskant JJ, de Jonge AJR (1985) Different subpopulations of haemocytes in juvenile, adult and *Trichobilharzia ocellata* infected *Lymnaea stagnalis*: a characterization using monoclonal antibodies. Z Parasitenkd 71: 815-819

Chapter 10 Transplantation Immunity in Arthropods: Is Immunorecognition Merely Wound-Healing?

A. M. LACKIE[1]

CONTENTS

1 Introduction

An animal such as an arthropod, whose means of support is carried externally to the soft tissues and whose blood circulates through large haemocoelic spaces which are in intimate contact with the cuticle, has a lot to lose if its cuticle is punctured and wounded. It is of prime importance that the wound is plugged immediately, in order to prevent haemolymph leakage and microbial invasion and, indirectly, to restore the integrity of the mechanical support system. As long as the surfaces lining the haemocoele are intact there is little or no adhesion of haemocytes to them; the few haemocytes which are observed adhering to the surfaces of tissues [1] are perhaps merely carrying out a house-keeping role in responding to small molecular irregularities, an argument which has also been used for the primary phagocytic cells of mammals, the neutrophil granulocyte. Any alteration in the tissue surface, whether by mutation [50] or by physical damage [55] will provoke a haemocytic "wound healing" response during which haemocytes aggregate around the altered region. Thus, any transplant or parasite which comes into contact with the haemolymph will be treated as damaged "self" and will stimulate a comparable reaction; the end result of the haemocytic encapsula-

1 Department of Zoology, The University, Glasgow G12 8QQ Scotland, U.K.

Immunity in Invertebrates
Ed. by M. Brehélin

tion or aggregation process, whether in response to wounds or to transplants, would be to produce a surface with properties similar to intact self (see Sect. 2.1.1). Such a system will be capable of different degrees of response in that more severe damage will produce a greater response and transplants whose surface properties differ widely from self will be more provocative. Thus a graded rather than an all-or-none type of haemocytic response would be expected, the extent of the response depending on the degree of the foreignness of the surface.

Transplantation immunology in insects – the assays, the results, and their interpretation – is illustrated and discussed from an immunologist's viewpoint in considerable detail elsewhere [34]. In this review, an alternative hypothesis will be examined, that an immune system is lacking in the arthropods and that all the reported transplantation phenomena are consistent with the operation of a system whose sole function is to respond to wounded self.

The ability of arthropods to recognize and mount a haemocytic response against transplanted cells and tissues will be reviewed before considering the mechanisms that would allow the insect response system to distinguish between intact and damaged self. The evidence for enhancement or re-education of the haemocytic response, and whether this is consistent with immunorecognition or a wound-healing response will be discussed in Sect. 4.

2 The Range of the Response

A variety of methods for assaying transplant recognition in arthropods have been developed and are discussed in detail elsewhere [34], but all rely on the haemocytic response – either as encapsulation or as aggregation (nodule formation) – providing an indication of whether or not the transplant has been recognized as foreign. The assays have included organ implantation, skin-grafting and haemolymph transfer, and the response to allogeneic (intraspecific) and xenogeneic (interspecific, intergeneric, etc.) transplants has been examined.

The important point to note, however, is that all tissue surfaces which normally abut the arthropodan haemocoele are covered by a basement membrane or connective tissue layer (Fig. 1) [2,3] and it is likely that an alteration in the properties of this layer will elicit a response [53].

2.1 Allogeneic Transplantation

2.1.1 Tissue Implantation

Experiments on tissue implantation fall into two categories; those in which the primary aim has been to look at the haemocytic response and those where the survival of the transplant can be deduced secondarily from the outcome of the manipulation. A variety of tissues have been implanted (Table 1), and have continued to develop or function normally. In the cases where this has been investigated, it is notable that hae-

Table 1. Allogeneic organ implantation in crustaceans and insects

Crustaceans	Tissue Transplanted	References
Carcinus	Sinus glands; "Y" organ tissue	[16,42]
Lysmata; Armadillidium	Ovaries	[8,38]
Asellus; Orchestia	Androgenic gland	[10]
Gammarus	Hepatic caeca	[12]
Insects[a]		
Drosophila, various cockroach spp.	Ovaries	[4,18,43,50]
Various cockroach spp.	Nerve cord	[21,28,37,58]
Drosophila	Imaginal discs	[6,18,50]

[a] A limited number of examples are listed here

mocytes do not adhere to the intact surface of the implant but attach only at its cut ends, where the connective tissue layer is damaged.

If abiotic material, such as a piece of nylon thread, is implanted into the haemocoele, it is rapidly encapsulated (see Chap. 12). The capsule eventually becomes covered with a layer of "coating material" [22] which may have properties in common with the connective tissue layer – both these layers stain with Alcian blue (Fig. 1) [36], indicating that they contain glycosaminoglycans [3]. The completed haemocytic capsules, transplanted allogeneically, do not stimulate further encapsulation [34,55] suggesting that once a capsule is completed it resembles intact self.

2.1.2 Cuticular Grafting

Skin-grafting – that is, orthotopic transplantation of cuticle and its underlying epidermis – appears to have been carried out only in the insects. The haemocytic response to transplanted adult cuticle is relatively slight and mainly confined to wound-healing around the edges of the graft [23], the haemocytic plug presumably forming a surface suitable for epithelial cell migration [6,11]. Allogeneic adult cuticle grafted onto adult cockroaches of the genus *Blaberus* or stick insects of the genus *Extatosoma* was not rejected, according to the criteria of melanization and cracking used by Thomas and Ratcliffe [63,64] and cuticle grafted onto adult *Periplaneta* remained in place and unmelanized until termination of the experiment (Lackie, unpubl.). However, due to the opacity of the cuticular exoskeleton, it is not always possible by examining the gross morphology of the graft, to ascertain the nature of the underlying haemocytic and epidermal events. If nymphal cuticle is transplanted onto nymphal recipients it is not necessary to make subjective measurements of the survival time because reappearance of donor-type cuticle at the graft-site after the recipient has moulted indicates that the donor epidermis has survived to produce new cuticle of its own type. Much of the early literature has been discussed by Wigglesworth [67] but some relevant examples of allotransplantation are the works of Piepho [45] on *Galleria* larvae and of Wigglesworth [67] and Locke [39] on *Rhodnius* nymphs. Pieces of limb in the cricket *Acheta* [51], the stick insect *Carausius* [60] and the cockroach *Leucophaea* [6] and even posterior abdominal segments of the cockroach *Periplaneta* [5] have been transplanted onto their respective allogeneic nymphs and the donor's subcuticular epidermis has fused with that of the recipient, producing new cuticle-covered appendages in synchrony with the moulting recipient. Similarly, in *Periplaneta*, nymphal allografts reversed through 180°, so that the origin of the cuticle at the graft-site could be distinguished post-moult by examining orientation of the scale pattern, were found to have healed in and the donor epidermis to have produced new cuticle at ecdysis [32].

Another way of investigating the response to allogeneic, nymphal, subcuticular epidermis is by heterotopic transplantation – implantation of a piece of cuticle into the abdominal haemocoele. If the allografted epidermis is treated as self, it persists and grows to form an "epidermal cyst" completely enclosing its own cuticle, and producing new cuticle within the cyst when the recipient moults. Such alloimplants survive and moult in a variety of insects including *Rhodnius* [67], *Manduca* [49] and *Periplaneta*.

2.1.3 Haemolymph Transfer

Insect haemocytes in suspension have been transferred allogeneically, usually in the course of experiments to determine whether immunity to soluble molecules [48], bacteria [65] or filarial nematode larvae (Ham P., pers. comm.) can be passively transferred. The fate of the transferred cells or of whole haemolymph was not investigated in these studies but it has been shown recently [33a] that allogeneic whole haemolymph transferred between adult *Periplaneta* does not stimulate nodule formation within the recipient's haemolymph. Nodules are aggregates of haemocytes formed in response to particulate material such as trypanosomes (Takle, pers. comm.), bacteria [46] and fungal spores [20a]. The failure of *Periplaneta* to respond to allogeneic transplants of haemocytes in these experiments is consistent with the idea that unaltered self will not provoke a response. A interhaemocytic response will not normally be stimulated within an individual unless the surfaces of some of the haemocytes become altered; this could occur by physical damage or through changes in the haemocytic plasma membrane and degranulation, which itself may be stimulated by contact with altered self [56].

2.2 Xenogeneic Transplantation

In general, implants of xenogeneic ovariole [4,27,28,35], imaginal disc [18,50] or ventral nerve cord [28,32,37,58] between insect species are encapsulated (Fig. 2). The same is true for amphipod crustaceans of the genus *Gammarus*: hepatic caeca transferred from *G. pulex* to *G. duebeni*, provoke an encapsulation response [12]. In addition, completed haemocytic capsules around nylon thread, transplanted from larvae of the moth *Diataraxia* to larvae of the beetle *Tenebrio*, stimulate further encapsulation in the recipient similar to that mounted against *Diataraxia* tissue [52].

With one recorded exception the recipient's response to insect xenogeneic cuticular grafts, whether between adults [19,23,64] or nymphs [32,39] parallels that towards xenogeneic organ implants. The exception is transplantation from *Nauphoeta* to *Periplaneta*, in which implanted tissues scarcely provoke a response but the epidermis of ortho- or heterotopically transplanted cuticle is encapsulated (see Sect. 4).

Whether or not a haemocytic response will occur can, to a limited extent, be predicted on the basic of phylogenetic closeness between donor and recipient species, xenografts between more closely related species being less likely to provoke a response ([4,24,50; see 34,53 for reviews]). Neither tissue implants [4,28,32] nor cuticular transplants [33] nor whole haemolymph [33a] transferred between the closely related cockroaches *Blatta* and *Periplaneta*, (same subfamily) stimulate a haemocytic response. However tissue implants [4], cuticular transplants [19,23,32] and cuticular implants [34] from *Blaberus* to *Periplaneta* (different subfamily) rapidly become encapsulated by the recipient's haemocytes.

It is particularly noteworthy that the extent of the haemocytic response to cuticular transplants, measured by counting the number of haemocyte nuclei along a transect taken at right angles to the graft, increases as phylogenetic separation between donor and recipient cockroach species increases [23]. However, failure to respond does not provide infallible evidence for taxonomic relatedness. The orthopteran *Schistocerca*, for example, does not recognize tissue or whole haemolymph transferred from the dictyopteran *Periplaneta*, but this proves to be due to its poor ability to distinguish between self and non-self [29]. Thus, the inadequate response of a recipient species may derive either from failure to recognize the graft or incompetence at responding to anything; this contention will be discussed in greater detail in Sect. 3.1.

2.3 Conclusions

Allografted tissue does not elicit a haemocytic response unless its connective tissue layer is damaged by physical means [12,50,53] or by mutation [50].

Most xenografts, however, provoke a strong haemocytic reaction and, as discussed above, layers of haemocytes accumulate around an implant or underneath a cuticular graft, the thickness of the layer probably being dependent on the degree of dissimilarity between self and the implant's surface. The final result of encapsulation is to present a surface which appears like self and whose "wounds" can thus be considered to be healed.

However, xenogeneic transplants from closely related species or genera do not, in general, provoke a response but remain intact, metabolically and physiologically active, and may even heal in and form continuity with the host [24,32]. In cases where fusion between donor and recipient tissue occurs, it would be very instructive – from the point of view of the "wound-healing" hypothesis – to determine whether any molecular alterations subsequently occur in the connective tissue layer of the donor to make it more closely similar to that of the recipient. Such information may also be helpful in solving the controversy as to whether haemocytes are themselves responsible for laying down connective tissue material [2,59,68].

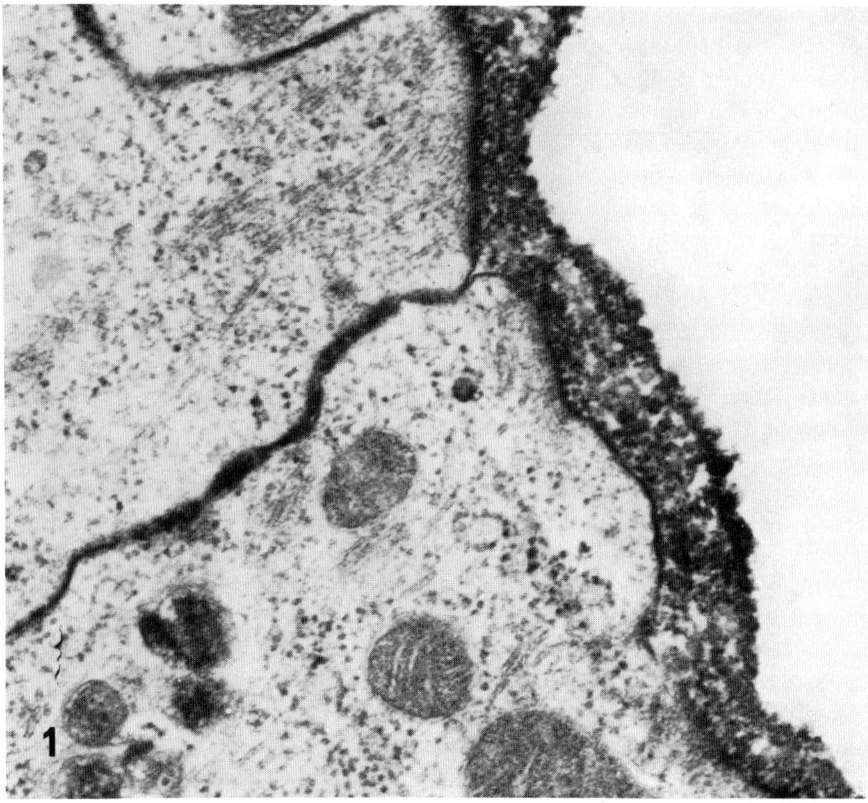
1

2

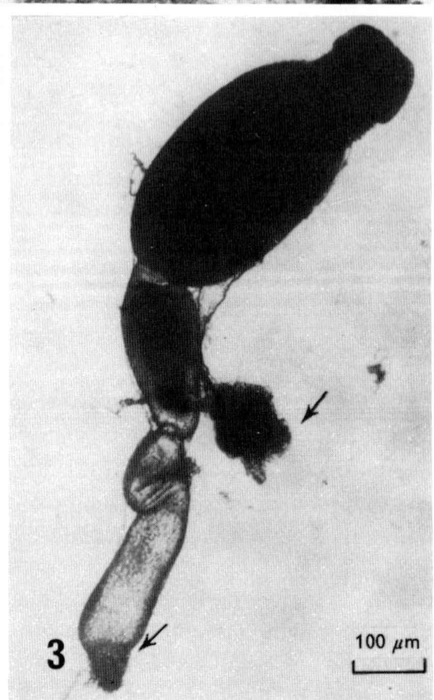
3
100 μm

3 How is Damaged Self Recognized?

When an insect is wounded, a transient haemocytopaenia occurs, accompanied by localized plasma clotting (see Chap. 14) and aggregation of haemocytes at the site of the wound [46]. The net result is that the wound is plugged and eventually restored to intact self. The extent of the response must be proportional to the size of the wound and must also, for obvious reasons, be self-limiting.

To understand how the magnitude of the response is regulated, it is first necessary to understand how altered self is recognized. Three mechanisms are possible: non-specific in response to alterations in the physicochemical surface properties of the connective tissue layer, specific via receptor-ligand interactions or, thirdly, alteration of the haemocytes themselves by release of so-called wounding factors.

3.1 Physicochemical Properties

Two physicochemical properties which might influence haemocyte adhesion are wettability and charge. This topic is comparatively well-researched for mammalian cells with respect to both in vitro adhesion to modified surfaces [14,20] and phagocytosis [17,40] and the role of cell and substratum surface charge in adhesion has been treated theoretically in a number of papers [13,14,44].

If an alteration in physicochemical properties is to trigger haemocytic adhesion, it must be assumed that the charge and wettability of all intact self surfaces fall within a particular range; the narrower the range, the more likely it is that minor alterations in a surface will push it outside the acceptable limits and trigger a response. The assumption that the surface properties of intact self comprise a fairly limited range is entirely reasonable for the arthropods, because all surfaces confronting the haemocoele are covered in a thin layer of connective tissue which contains collagen and proteoglycans [3]. The latter class of molecules are characteristically polyanionic at physiological pH, the negative charge being imparted predominantly by sulphated glycosaminoglycans (GAGs) attached to the protein backbone of the proteoglycan.

Whether the composition, and thus the charge and wettability, of the connective tissue varies according to the tissue it coats has not been studied, although there is histochemical and cytological evidence to suggest that the neural lamina may be rather different from the other surfaces [3,59]. Such differences in composition may explain why nerve cord of the cockroach *Nauphoeta* stimulates very little response in *Peripla-*

Fig. 1. The connective tissue layer (basement membrane) underlying the subcuticular epidermis of *Periplaneta*. The tissue was stained with Alcian blue at the fixation stage; the stain reacts with glycosaminoglycans in the basement membrane to form granular deposits

Fig. 2. Nerve ganglion and cord 24 h after implantation into a xenogeneic recipient in which it was recognized as differing from intact self and was completely encapsulated by haemocytes

Fig. 3. Ovariole 24 h after implantation into a xenogeneic recipient, in which it was not recognized as differing from intact self. The recipient's haemocytes have only responded to the damaged ends of the implant (*arrows*)

neta, although cuticular grafts and implants from the same species are rapidly coated with haemocytes [32,34].

That differences in charge and wettability are important in recognition has been shown by experiments in which abiotic objects have been implanted into the insect haemocoele and the degree of haemocytic encapsulation measured [15,33,66]. Haemocytes of *Periplaneta* adhere to positive, neutral and negative foreign surfaces in vivo, the capsule thickness being dependent on the charge of the object, but although *Schistocerca* haemocytes can encapsulate neutral and positive surfaces, negatively charged surfaces provoke no response and remain unencapsulated [33]. By investigating the haemocytic response to carboxymethylated agarose beads of different negativities, it has been found that encapsulation is not stimulated until the surface charge of the bead is close to neutrality [33b]. Since most biotic surfaces including those of connective tissue-coated organ transplants are negatively charged, these observations may help to explain why locusts are so poor at recognizing and responding to xenografts [27,28,35].

Using two different methods, cell electrophoresis of living haemocytes in suspension and measurement of bound cationized ferritin, Takle [62] has shown that the surface negativity of locust haemocytes (zeta potential of -26.6 ± 6.7 mV) is significantly greater than that of *Periplaneta* haemocytes (-12.8 ± 5.8 mV). There is thus the possibility that by virtue of their strong negativity, locust haemocytes are unable to approach and adhere to negatively charged substrata; cockroach cells would have a greater range of interactions available to them. Certainly, the in vivo responsiveness of the haemocytes of the two species would accord well with this view. Despite these deficiencies, however, the haemocytes of *Schistocerca* are entirely capable of mounting an effective wound-healing response.

3.2 Specific Receptor-Ligand Interactions

Serum lectins have long been regarded as the likely candidates for mediators of specific immunorecognition in arthropods, but so far there is only circumstantial evidence in favour of this role [41]. An alternative role would be their involvement in wound-healing and morphogenesis, as has been suggested by Komano et al. [26]. These authors have shown that wounding *Sarcophaga* larvae stimulates the production of a galactose-binding lectin, which they suggest is responsible for recognition of "disintegrating tissue fragments" since it is also induced at the time of internal rearrangement within the metamorphosing pupa.

The carbohydrate specificity of lectins might allow them to recognize differences in the connective tissue layer covering organs. Damage might expose hidden carbohydrate epitopes of the constituent GAGs and glycoproteins. It has been argued that lectin specificity for sialic acid, for example, is probably inappropriate in immunological defence against microorganisms [41]; such a specificity might be more appropriate for recognizing damaged self in that damage to the connective-tissue layer might expose sialic acid groups on the underlying cell membrane. Soluble lectins could then bind and "mark" the surface for subsequent haemocyte adhesion, or membrane-bound lectins at the haemocyte surface might bind directly. Haemocyte-lectin interaction could promote adhesion and could also stimulate the release of haemocytic factors which could recruit additional haemocytes to the capsule [61], promote epithelial cell migration and wound-healing [7], or which could be incorporated into connective tissue [68].

Until purified lectins have been obtained and their range of biological reactivity compared with the range of host responsiveness to transplants, no definite conclusions can be drawn about their role in recognition of transplant surfaces. It is, however, interesting that the agglutinating activity of *Schistocerca* serum has a smaller range and titre than that of *Periplaneta* and that the former species is much inferior at recognizing the difference between intact self and transplants [29,30,31].

3.3 Conclusions

Although the extent of the haemocytic response to foreign surfaces is quite clearly influenced by their physicochemical properties, the rôle of surface carbohydrates and lectins can at present only be speculated upon. From the available evidence it is difficult to decide whether these parameters are involved in recognizing damaged self or in immunological recognition of non-self.

4 Enhancement and Re-Education of the Haemocytic Response

It is comparatively easy to understand how the haemocytic response, whether immunological or reparative, could undergo short-term enhancement; the process is analogous to kicking a hive of bees twice, in that the first kick would so infuriate the colony that its response to the second would be massive and instantaneous. Although there is ample evidence of an enhanced response after immunization against micro-organisms (see Boman, Chap. 6) and soluble molecules [48], there is little evidence in favour of an increased rate of encapsulation of transplanted tissues. Encapsulation of implants occurs so rapidly [56] that enhancement after challenge would probably be difficult to quantify but it should be possible to measure the extent of accumulation of haemocytes underneath a "second-set" graft using the method of Karp and coworkers [19].

The epidermis of nymphal *Blatta* cuticle grafted onto nymphal *Periplaneta* heals in and produces new *Blatta*-like cuticle when the recipient moults [32]. Injection of pieces of *Blatta* nerve cord into recipients prior to grafting did not lead to rejection of the cuticular graft [32] but this is probably not surprising, since these implants would not, in any case, have been encapsulated ([28]: and Sect. 2.2) – in other words, prior stimulation of the recipient's haemocytes would only have been minimal, in response to the small puncture wound.

Throughout this discussion, it has been assumed that recognition of a surface as either damaged self or non-self will depend on its deviation from the acceptable range of variation in surface properties. In the locust, as argued elsewhere [29], this range is broad and most foreign surfaces fall within its upper and lower limits and are treated as intact self; damaged self is obviously excluded because wounds are healed. In the cockroach, since all foreign surfaces tested [28,32,33] – with the exception of *Blatta* tissues and certain habitual parasites [35,37] – provoke an encapsulation response, the range is narrower, allowing the insect to maintain its integrity more successfully.

Wether or not an arthropod can be re-educated so that its haemocytes respond to a foreign surface which is normally within the acceptable range will depend on whether the limits beyond which recognition occurs can be re-set, or whether the surface itself can be pushed beyond the limits. The latter could occur, relatively slowly, by physiological damage due to being in the wrong environment or it could occur instantaneously due to minor but nevertheless notable differences in the molecular or physicochemical properties of the surfaces being presented. The equivocal response to *Nauphoeta* nerve cord by *Periplaneta* haemocytes may be due to this tissue lying at the limit of recognition; difference is only perceived when the *Nauphoeta* surface presented is the "basement membrane" underlying cuticular grafts [32,34]. In this example, no manipulation of the acceptable range occurs, it is the method of presentation of the surface and the possible differences between them (Sect. 3.1) which are important.

Preliminary evidence from our laboratory indicates that the limits of recognition in *Periplaneta* can be manipulated so that the insect can be persuaded to respond to *Blatta* cuticular grafts. It was noted above that prior implantation of *Blatta* nerve cord did not immunize the recipient *Periplaneta* against subsequent cuticular grafts, presumably because the implants did not themselves become encapsulated. In the series of experiments shown in Table 2, nymphal *Periplaneta* were injected with pieces of *Blaberus* cuticle, which are known to be encapsulated [32], and then groups of recipients were grafted with nymphal *Blatta* cuticle at 24 h intervals thereafter. The results show that on those insects given *Blatta* grafts from 24–72 h after the final implantation of *Blaberus* cuticle, no *Blatta*-type cuticle reappeared post-moult, indicating that the "basement membrane" of the graft's subcuticular epidermis had stimulated a haemocytic response and had been replaced by *Periplaneta* epidermis. Control insects, implanted with *Blatta* cuticle, accepted all the *Blatta* grafts as self as in previous studies. The most likely explanation is that by activating *Periplaneta's* haemocytes with the *Blaberus* implants, the range within which xenografts are accepted as intact self has temporarily been reduced so that *Blatta* is now excluded.

Reducing the acceptable range – "re-educating" the response mechanism – could be accomplished by (a) changing the surface properties of the responsive cells or (b) increasing the plasma concentration of molecules which could bind specifically or non-specifically to the foreign surface. In the latter case – for example, release of specific lectins or release of putative by-products of the proPOA pathway (see Chap. 15) – merely increasing the concentration will not be sufficient. Two further assumptions must be made: that even in the unstimulated recipient the transplant binds a small number of molecules, too few however to promote haemocyte adhesion, and that there is differential binding to recipient and transplant surfaces. Without differential binding, there would be no reason why both self and transplant surface should not be pushed beyond the limits of acceptance in the stimulated host, with disastrous consequences. Both these assumptions have the advantage that they can be tested as techniques improve.

Table 2. The effect of prior implantation of *Blaberus* tissue into nymphal *Periplaneta* on subsequent survival of *Blatta* cuticular xenografts

	Hours post-implantation at which *Periplaneta* nymphs received *Blatta* grafts				
No. recipients bearing *Blatta* type cuticle post-moult/No. grafted (%)	24	48	72	96	120
Experimental[a]	2/9 (22)	0/7 (0)	2/8 (25)	8/8 (100)	10/10 (100)
Control[b]	9/9 (100)	10/10 (100)	10/10 (100)	7/7 (100)	5/5 (100)

[a] Experimental group implanted with *Blaberus* tissue at days −8 and 0
[b] Control group implanted with *Blatta* tissue at days −8 and 0

Re-education occurring through alterations in the surface properties of the activated haemocyte can also be tested. Such alterations may occur by expression of new membrane components so that either the resultant changes in surface charge allow the cell to adhere non-specifically to the foreign surface or increased expression of specific receptors encourages receptor-ligand interaction.

Whatever the mechanism of this re-educative phenomenon, its result – the lowering of the recognition threshold – can at present be equally well explained in terms of either immunorecognition or wound-healing.

5 General Conclusions

The haemocytic response of arthropods is a system which can recognize air bubbles inadvertently introduced during tissue implantation and can distinguish the degree of hydrophobicity or surface charge of a foreign surface. The system cannot distinguish between intact self, intact allografts, some intact xenografts or even intact habitual parasites [53,55], yet it will respond to wounds in all of these (see Figs. 1-3) and can be persuaded to respond to a previously unrecognized intact surface. Mechanisms by which haemocytic recognition and re-education could occur have been suggested and discussed. However, from the evidence available, it is not possible to decide whether the haemocytic response is triggered by immunological recognition of non-self or as a wound-healing response to damage self. By discussing the response in terms of wound-healing, I have tried to focus attention on some of the enigmas which confront us and hope that various areas for future research, which would allow us to to distinguish between these hypotheses, might be seen a little more clearly. With our improved ability to manipulate the humoral and haemocytic components of the arthropodan defence system and with the increasingly sophisticated immunochemical and biochemical techniques available, it is now becoming possible to devise and perform the necessary experiments to test the various hypotheses. Insect immunology is entering a rapid growth phase.

Acknowledgements. Thanks to Garry Takle and Bubbly Dularay for allowing me to quote unpublished data; to John Lackie for helpful discussions; to Lawrence Tetley for the electron microscopy of basement membranes, and to the SERC for grants GRC 01351 and GRC 87959.

References

1. Arnold JW (1959) Observations on living haemocytes in wing veins of the cockroach *Blaberus giganteus*. Ann Entomol Soc Am 52: 229-236
2. Ashhurst DE (1979) Haemocytes and connective tissue. In: Gupta AP (ed) Insect haemocytes. Cambridge University Press, Cambridge, London
3. Ashhurst DE (1982) The structure and development of insect connective tissues. In: King S, Akai H (eds) Insect ultrastructure, Vol 1. Plenum, New York
4. Bell WJ (1972) Yolk formation by transplanted cockroach oocytes. J Exp Zool 181: 41-48
5. Bodenstein D (1953) Studies on humoral mechanisms in the growth and metamorphosis of the cockroach *Periplaneta americana*. J Exp Zool 123: 189-232
6. Bohn H (1972) The origin of the epidermis in supernumerary regenerates of triple legs in cockroaches (Blattaria). J Embryol Exp Morphol 28: 185-208
7. Bohn H (1975) Growth-promoting effects of haemocytes on insect epidermis in vitro. J Insect Physiol 21: 1283-1293
8. Carlisle DB (1953) Studies on *Lysmata seticaudata* Risso (Crustacea Decapoda) V. The ovarian inhibiting hormone and the hormonal inhibition of sex reversal. Pubbl Staz Zool Napoli 24: 355-372
9. Chain BM, Anderson RS (1982) Selective depletion of plasmatocytes in *Galleria* following injection of bacteria. J Insect Physiol 28: 377-384
10. Charniaux-Cotton H (1960) Sex determination. In: Watermann TH (ed) Physiology of crustacea, vol 1. Metabolism and growth. Academic Press, New York
11. Clarke RM, Harvey WR (1965) Cellular membrane formation by plasmatocytes of diapausing cecropia pupae. J Insect Physiol 11: 161-169
12. Crompton DWT (1967) Studies on the haemocytic reaction of *Gammarus* species and its relationship to *Polymorphus minutus* (Acanthocephala). Parasitology 57: 389-401
13. Curtis ASG (1972) Adhesive interactions between organisms. In: Taylor AER, Muller R (eds) Functional aspects of parasite surfaces. BSP Symposium, vol 10. Blackwell, Oxford, pp 1-18
14. Curtis ASG (1984) Cell adhesion. In: Graham CF, Wareing PF (eds) Developmental control in animals. Blackwell, Oxford, pp 98-119
15. Dunphy GB, Nolan RA (1982) Cellular immune response of spruce budworm larvae to *Entomophthora egressa* protoplasts and other test particles. J Invertebr Pathol 39: 81-92
16. Echalier G (1955) Rôle de l'organe Y dans la déterminisme de la mue de *Carcinides (Carcinus) moenas* (L) (Crustaces: Decapodes). Expériences d'implantation. CR Acad Sci Paris D 240: 1581-1583
17. Edebo L, Kihlstrom E, Magnusson KE, Stendahl O (1980) The hydrophobic effect and charge effects in the adhesion of enterobacteria to animal cell surfaces and the influences of antibodies of different immunoglobulin classes. In: Curtis ASG, Pitts JD (eds) Cell adhesion and motility. BSCB Symposium, vol 3. Cambridge University Press, Cambridge, pp 65-101
18. Ephrussi B, Beadle GW (1936) A technique for transplantation for *Drosophila*. Am Nat 70: 218-225
19. George JF, Karp RD, Rheins LA (1984) Primary integumentary xenograft reactivity in the American cockroach *Periplaneta americana*. Transplantation (Baltimore) 37: 478-484

20. Gingell D, Vince SA (1982) Substratum wettability and charge influence the spreading of *Dictyostelium* amoebae and the formation of ultrathin cytoplasmic lamellae. J Cell Sci 54: 255-285
20a. Gunnarsson S, Lackie AM (1985) Haemocytic aggregation in the locust *Schistocerca gregaria* and the cockroach *Periplaneta americana* in response to injected molecules of microbial origin. J Invertebr Pathol (in press)
21. Guthrie DM (1966) Physiological competition between host and implanted ganglia in an insect (*Periplaneta americana*). Nature 210: 312-313
22. Hillen ND (1977) Experimental studies on the reactions of insect haemocytes to artificial implants and habitual parasitoids, and on the initiation of wound-healing in insects. Ph D Thesis, University of London
23. Jones SE, Bell WJ (1982) Cell-mediated immune-type response of the American cockroach. Dev Comp Immunol 6: 35-42
24. Kambysellis MP (1970) Compatibility in insect tissue transplantations. J Exp Zool 175: 169-180
25. Karp RD, Rheins LA (1980) Induction of specific humoral immunity to soluble proteins in the American cockroach II. Nature of the secondary response. Dev Comp Immunol 4: 629-637
26. Komano H, Mizuno D, Natori S (1981) A possible mechanism of induction of insect lectin. J Biol Chem 256: 7087-7089
27. Lackie AM (1976) Evasion of the haemocytic defence reaction of insects by larvae of *Hymenolepis diminuta* (Cestoda). Parasitology 73: 97-104
28. Lackie AM (1979) Cellular recognition of foreign-ness in 2 insect species, the American cockroach and the desert locust. Immunology 36: 909-1007
29. Lackie AM (1981a) Immune recognition in insects. Dev Comp Immunol 5: 191-204
30. Lackie AM (1981b) The specificity of the serum agglutinins of *Periplaneta americana* and *Schistocerca gregaria* and its relationship to the insects' immune response. J Insect Physiol 27: 139-143
31. Lackie AM (1981c) Humoral mechanisms in the immune response of insects to larvae of *Hymenolepis diminuta* (Cestoda). Parasite Immunol (Oxf) 3: 201-208
32. Lackie AM (1983a) Immunological recognition of cuticular transplants in insects. Dev Comp Immunol 7: 41-50
33. Lackie AM (1983b) Effect of substratum wettability and charge on adhesion in vitro and encapsulation in vivo by insect haemocytes. J Cell Scie 63: 181-190
33a. Lackie AM (1985a) Haemolymph transfer as an assay for immunorecognition in insects. Transplantation (Baltimore) (in press)
33b. Lackie AM (1985b) The role of substratum surface charge in adhesion and encapsulation by locust haemocytes in vivo. J Invertebr Pathol (in press)
34. Lackie AM (1986) Transplantation immunology in insects: the limits of recognition. In: Gupta AP (ed) Cellular and humoral immunity in arthropods. Wiley & Sons, New York, London (in press)
35. Lackie AM, Lackie JM (1979) Evasion of the insect immune response by *Moniliformis dubius* (Acanthocephala): further observations on the origin of the envelope. Parasitology 79: 297-303
36. Lackie AM, Takle GB, Tetley L (1985) Haemocytic encapsulation in the locust *Schistocerca gregaria* (Orthoptera) and the cockroach *Periplaneta americana* (Dictyoptera). Cell Tissue Res 240: 343-351
37. Lackie JM (1975) Host-specificity of *Moniliformis dubius*, a parasite of cockroaches. Int J Parasitol 5: 301-309
38. Legrand JJ (1955) Rôle endocrinien de l'ovaire dans la différenciation des oostégites chez les Crustaces Isopodes terrestres. CR Acad Sci Paris D 241: 1083-85
39. Locke M)1974) Structure and formation of the integument in insects. In: Rockstein M (ed) Physiology of the Insects, vol 6. Academic Press, New York, pp 129-221
40. Magnusson KE, Davies J, Grundstrom T, Kihlstrom E, Normak S (1980) Surface charge and hydrophobicity of *Salmonella*. *E. coli* and *Gonococci* in relation to their tendency to associate with animal cells. Scand J Infect Dis Suppl 24: 135-140
41. McSweegan EF, Pistole TG (1982) Interaction of the lectin limulin with capsular polysaccharides. Biochem Biophys Res Commun 106: 1390-1397

42. Passano LM (1953) Neurosecretory control of moulting in crabs. Physiol Comp 3: 155-189
43. Peterson L (1968) Cellular immune responses of insects to foreign tissue implants. Ph D Thesis University of Illinois
44. Pethica BA (1980) Microbial and cell adhesion. In: Berkeley RCW, Lynch JM, Melling J, Rutter PR, Vincent B (eds) Microbial adhesion to Surfaces. Horwood, Chichester, pp 19-45
45. Piepho H (1943) Wirkstoffe in der Metamorphose von Schmetterlingen und anderen Insekten. Naturwissenschaften 31: 329-335
46. Ratcliffe NA, Rowley AF (1979) Role of hemocytes in defence against biological agents. In: Gupta AP (ed) Insect hemocytes. Cambridge University Press, Cambridge, pp 331-414
47. Ratcliffe NA, Leonard C, Rowley AF (1984) Prophenoloxidase activation: nonself recognition and cell cooperation in insect immunity. Science 226: 557-559
48. Rheins LA, Karp RD (1984) The humoral immune response in the American cockroach *Periplaneta americana*; reactivity to a defined antigen from honeybee venom, phospholipase A_2. Dev Comp Immunol 8: 791-801
49. Riddiford L (1978) Ecdysone-induced change in cellular commitment of the epidermis of the tobacco hornworm *Manduca sexta*, at the initiation of metamorphosis. Gen Comp Endocrinol 34: 438-446
50. Rizki RM, Rizki TM (1980) Hemocyte responses to implanted tissues in *Drosophila melanogaster* larvae. Wilhelm Roux's Archives 189: 207-213
51. Sahota TS, Edwards JS (1969) Development of grafted supernumerary legs in the house-cricket *Acheta domesticus*. J Insect Physiol 15: 1367-1373
52. Salt G (1960) Experimental studies in insect parasitism XI. The haemocytic reaction of a caterpillar under varied conditions. Proc Roy Soc Lond B Biol Sci 151: 446-457
53. Salt G (1961) The haemocytic reaction of insects to foreign bodies. In: Ramsay JA, Wigglesworth VB (eds) The cell and the organism. Cambridge University Press, Cambridge, pp 175-192
54. Salt G (1963) Experimental studies in insect parasitism XII. The reactions of 6 exopterygote insects to an alien parasite. J Insect Physiol 9: 647-667
55. Salt G (170) The cellular defence reactions of insects. Cambridge University Press, Cambridge
56. Schmit AR, Ratcliffe NA (1976) The encapsulation of foreign tissue implants in *Galleria mellonella* larvae. J Insect Physiol 23: 175-180
57. Schubiger G (1971) Regeneration, duplication and transdetermination in fragments of the leg disc of *Drosophila melanogaster*. Dev Biol 26: 277-295
58. Scott M (1971) Recognition of foreign-ness in invertebrates: transplantation studies using the American cockroach *Periplaneta americana*. Transplantation (Baltimore) 11: 78-85
59. Shrivastava SC, Richards AG (1965) An autoradiographic study of the relation between haemocytes and connective tissue in the waxmoth *Galleria mellonella*. Biol Bull (Woods Hole) 128: 337-345
60. Steinberg DM (1959) Regeneration of homografted and heterografted limbs in the stick insects (Phasmodea). Dokl Acad Sci USSR Biol Sci 129: 1001-1003
61. Takle GB (1985) The effect of components of the prophenoloxidase-activation pathway on the chemokinetic locomotion of insect haemocytes in vitro. Proc Invertebr Immunol Conf Montpellier 1984. Dev Comp Immunol 9: 164
62. Takle GB, Lackie AM (1985) Surface-charge of insect haemocytes measured by cell electrophoresis and cationised ferritin-binding. J Cell Sci 75: 207-214
63. Thomas IG, Ratcliffe NA (1980) Studies on recognition of foreign-ness in insects utilising integumental transplants. In: Solomon JB (ed) Aspects of developmental and comparative immunology, Pergamon, New York, pp 105-110
64. Thomas IG, Ratcliffe NA (1982) Integumental grafting and immunorecognition in insects. Dev Comp Immunol 6: 643-654
65. de Verno PJ, Aston WP, Chadwick JS (1983) Transfer of immunity against *Pseudomonas aeruginosa* P11-1 in *Galleria* larvae. Dev Comp Immunol 7: 423-634
66. Vinson SB (1974) The role of the foreign surface and female parasitoid secretions on the immune response of an insect. Parasitology 79: 297-306
67. Wigglesworth VB (1954) The physiology of insect metamorphosis. Cambridge University Press, Cambridge
68. Wigglesworth VB (1979) Haemocytes and growth in insects. In: Gupta AP (ed) Insect hemocytes. Cambridge University Press, Cambridge

Chapter 11 Evolution of Histoincompatibility

E. L. COOPER[1]

CONTENTS

1 The Earthworm's Immune System?

1.1 Introduction

Earthworms have been known in modern times since Darwin, but in relation to improvement of soil [23]. Comparative immunologists interested in evolution of immunity have now rediscovered the earthworm [11,12,29,31,40,43,71] to search for origins of immune mechanisms [13,14], as did Metchnikoff's crucial work and that of other early comparative immunologists [46], since certain invertebrates, including earthworms, may possess cells from which higher immune systems evolved. The immune system of animals in the phylum, Annelida, (e.g., the earthworm) is housed in the coelom, which contains coelomic fluid, and coelomocytes, the worm's leukocytes. Coelomocytes, like leukocytes, are sensitive to infections and are active in defense reac-

1 Department of Anatomy, Center for Health Sciences, University of California, Los Angeles, California 90024, USA

Immunity in Invertebrates
Ed. by M. Brehélin

tions ranging from phagocytosis to the more complex cellular responses that govern the mechanisms of tissue graft rejection and those that ensure the synthesis of humoral defense substances. To react against foreign material, coelomocytes surely possess cellular recognition units or receptors, whose nature is still unknown. The receptors for antigens in mammals are antibodies, but antibodies or immunoglobulins have not been found in any invertebrates. Such species must, therefore, have different and perhaps simpler, primordial receptor units for antigen which may prove to be members of the Ig superfamily. The fundamental problem in studies of invertebrates in general using the earthworm's immune system as an example, is to explain the process of receptor-mediated recognition of antigen.

1.2 What Characterizes the Earthworm's Immune System?

Are the immune response of invertebrates related to these more complex ones? Some comparative immunologists once suggested that graft rejection capacities of earthworms are mediated by T-cell progenitors, whereas other workers have disagreed with this point [28,66]. Phagocytosis is the defense reaction common to all animals, even the simplest, the unicellular protozoans. The coelomic cavity of earthworms contain numerous coelomocytes which search out foreign material, phagocytose and destroy it [58]. Earthworms like other animals possess an efficient, *nonspecific* mechanism for disposing of foreign *non-self* material, which is as efficient as equivalent responses in humans. The *specific* immune system of earthworms, in contrast, also possesses certain characteristics common to the immune systems of vertebrates, but it may be less complex [16]. There are antigen-recognizing cells and effector cells. Antigen-recognizing cells may be able to sense antigen via the, as yet undemonstrated, cell surface receptors. Presumably these putative antigen recognition cells later interact with effector cells, a process which is responsible for immune functions, such as transplant rejection and agglutinin synthesis.

2 Transplantation of Autografts, Allografts, and Xenografts

2.1 Basic Characteristics

Earthworms have been important in studies concerning analyses of cell-mediated immunity, since they will accept grafts of *self*-tissue, but will recognize as different grafts of *non-self* tissue and reject it. All grafts, both *self* (autografts) and *non-self* (allografts, xenografts) heal initially, become vascularized and firmly attached to the host. Foreign grafts later show varying degrees of rejection, whereas self-tissue grafts remain permanently healed. Intrafamilial xenografts are rejected more vigorously and usually completely, whereas allografts are rejected more slowly and often incompletely [6-10,20, 25-27,32,39,51,60,62]. The earliest phase of graft rejection, the recognition of foreign tissue antigens, has been subjected to analysis by confronting host leukocytes with foreign erythrocytes [18]. Rabbit and rat erythrocytes prevented healing of allografts when grafts were transplanted and erythrocytes injected simultaneously. In contrast,

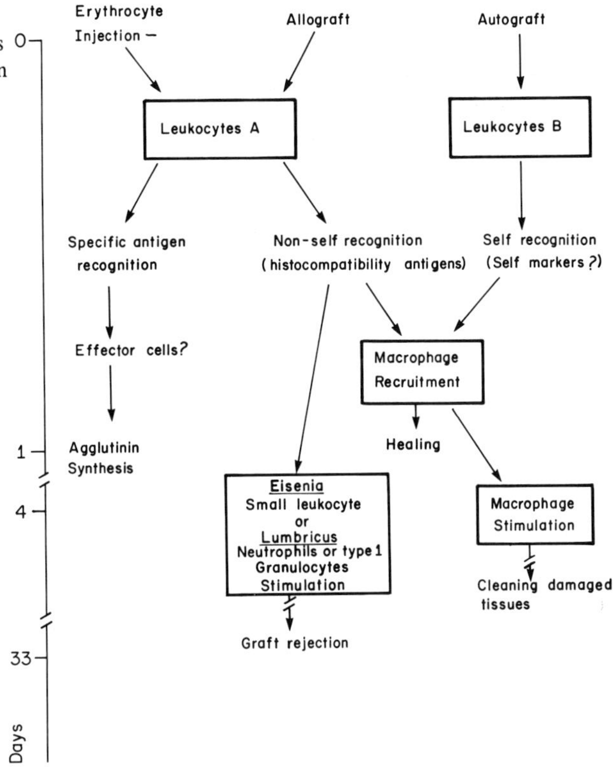

Fig. 1. Hypothetical scheme of earthworm leukocyte interactions occuring in grafting and agglutinin synthesis following erythrocytic injection, allografting, or autografting. (Note that graft rejection occurred at 33 days, whereas cleaning damaged tissues after day 33, not before) (From Cooper and Roch [18])

autografts and allografts transplanted on worms injected 1 or 2 days before grafting were never affected. Since earthworms readily produce higher titers of erythrocyte agglutinins at 24 h postinjection than at later times, it has been proposed that earthworm leukocytes interact during early phases of graft healing and agglutinin synthesis (Fig. 1).

2.2 Is Accelerated Rejection of Allografts and Intrafamilial Xenografts One Criterion of Immune Memory?

Second-set xenografts from *Eisenia foetida* transplanted onto *Lumbricus terrestris* hosts, will undergo accelerated rejection if grafted 1 week after grafting first-set xenografts. In contrast, second-set xenografts transplanted at longer intervals after first-set grafting, undergo no accelerated rejection. The essential variable in these experiments has been the time of second-set transplantation. Accelerated rejection as a measure of short-term memory, is important in earthworms and other invertebrates, and probably results from intense responses (although of short duration) of activated coelomocytes (Cooper E.L., in press, [17]). Second-set *Lumbricus terrestris* allografts from the same donor to the same recipient undergo accelerated rejection if transplanted less than 10 days after first-set allografts, but if transplanted at longer time intervals (20–90 days), second-set allografts do not show accelerated rejection. Accelerated rejection of *allo*transplants under these kinetic conditions agrees with earlier intrafamilial *xeno*transplanr rejection patterns stressing an essential experimental variable: the time of second-set transplantation. Thus, accelerated rejection is apparently one measure of the short lived "memory" 10-days-old component of the cellular defense immune system and probably results from intense responses (although of short duration) of coelomocytes activated by a first transplant still present at the time of second-set grafting, but absent or inactive at later periods [19].

2.3 What is the Role of Coelomocytes in Graft Destruction?

2.3.1 Transplant Destruction: Histopathology

Coelomocytes first accumulate near the graft sites of both autografts, allo- and xenografts and infiltrate the graft matrix within 24 h after transplanting grafts. Since infiltration ceases in autografts after damaged tissue has been removed, and healing is well underway, this first reaction is apparently a nonspecific one in response to injury [9,10,32]. Against xenografts, coelomocytes may completely wall off entire grafts in an encapsulation-like process [32,51], which may be related to the extent of injury to the graft bed [32]. By electron microscopy, within 1 to 3 days posttransplantation, the graft matrix is infiltrated by coelomocytes known as type-I granulocytes (neutrophils, by light microscopy), which destroy the viable inner muscle layer [39]. At day 5, type-I granulocytes contain numerous phagosomes, residual bodies, and glycogen granules and by 11 to 13 days, type-I granulocytes have partially destroyed the outer muscle layer, leaving the outer epithelium mostly intact. Small lymphocytic (basophilic) coelomocytes are often visible within grafts and they seem to function primarily as minor scavengers, phagocytosing small cellular debris, but not viable muscle fragments.

Valembois presents a somewhat different interpretation in xenograft experiments, employing *E. foetida* as host [61,64]. In *E. foetida*, destruction of first-set grafts apparently occurs as a result of autolysis, involving synthesis of lysosomal enzymes by the transplanted tissues. In contrast, destruction of second-set grafts results from coelomocyte invasion, as in first-set xenografts of *L. terrestris*. Valembois has identified the coelomocytes which infiltrate the graft as macrophages (type-I granulocytes). Coelomocytes also infiltrate autografts and wounds of sham-operated worms as a nonspecific, general inflammatory response, related to wound-healing and removal of damaged tissue, without visible phagocytosis of viable muscle [39].

2.3.2 Kinetics of Coelomocyte Response to Transplants

Grafting or injury to the earthworm's body wall is accompanied by increased coelomocytes, facilitating their quantitation by measuring numbers directly under autografts, xenografts, and wounds at different times during rejection (or, in the case of autografts or wounds, healing). Using *E. foetida* xenografts on *L. terrestris* hosts, coelomocyte responses to wounds rise most rapidly within 24 h, decline rapidly and return to normal by 72 h [33]. Autografts elicit a weaker coelomocytic response, rising only slightly above those of ungrafted worms and returning to normal by day 3. In contrast, coelomocytic responses to xenografts are slower, reach peaks in 3 to 4 days, decline more slowly, and require 7 days for coelomocytes to return to normal levels.

Xenografts destruction was complete in a mean time of 17 days at 20° C, approximately 10 day following the decline in coelomocyte numbers beneath transplants. After transplanting second grafts at this time (17 days posttransplantation of first xenograft) rejection occurred in an accelerated mean time of 6–7 days, indicating a memory component of the rejection response [33]. Quantitation of coelomocytes taken from second grafts or wounds varied considerably from those associated with first grafts or wounds. After second xenografts, coelomocyte numbers were 20–30% higher than after first grafts, peaking 1–2 days posttransplantation. Second transplants (including autografts) and wounds performed 5 days after the initial grafting, produced rejection times similar to those of the 17-day group. These accelerated and heightened responses against a second graft, suggest that a memory component is inducible in earthworm coelomocytes, but regulated by temperature [21].

2.3.3 Adoptive Transfer of Short-Term Memory

The role of coelomocytes in graft rejection is based upon adoptive transfer experiments [1,26,63]. Host *L. terrestris* (A) were first xenografted with *E. foetida*, then coelomocytes harvested at 5 day posttransplantation and injected into ungrafted *L. terrestris* (B). This second host was then xenografted with the same *E. foetida* donor used to induce immunity in A. Because *L. terrestris* shows only negligible allograft responses late after transplantation, [9] no strong coelomocytic allo-incompatibility was expected prior to the action of primed coelomocytes against the *E. foetida* graft transplanted to a *L. terrestris* host (B). The second host (B) showed accelerated rejection of its first transplant, demonstrating short-term memory and confirming that primed coelomocytes can adoptive transfer since coelomic fluid alone, free of coelomocytes, is ineffective. Moreover, coelomocytes from unprimed *L. terrestris* or from *L. terrestris* primed with saline are also unable to transfer the response. In this adoptive transfer experiment, we must define the cell type which effects the response.

3 How do Coelomocytes Respond to Tissue Transplant Antigens in Vivo?

In addition to mitogens, coelomocytes are also stimulated by putative transplantation antigens. Lemmi demonstrated the in vivo synthesis of DNA, measured by ^{3}H-thymidine incorporation into coelomocytes of xenografted *L. terrestris* at various times postgrafting, with a peak incorporation at 4 days [37]. Roch et al. investigated in vitro DNA synthesis by *E. foetida* coelomocytes following wounding, allografting, and xenografting [53]. Peak incorporation of ^{3}H-thymidine occurred 4 days later, but showed decreasing stimulation indices (in brackets): allografts [15] → xenografts [5] → wounds [4]. Responses following second grafts or wounds were more complex, with a peak at day 2 for xenografts [4], at days 2 and 6 for allografts [4], and at day 6 for wounds [3]. Xenografting (particularly second sets) stimated DNA synthesis in greater numbers of small coelomocytes (< 10 μm) than in larger cells (> 10 μm).

Stimulation of coelomocytes by both wounding and grafting is greater than that induced by mitogens [65]. Four days after xenografting, coelomocytes (13%) incorporated ^{3}H-thymidine in vitro; after second xenografts less (6%) were labelled. After first wounds 6.5% were labelled, and after second wounds only 4%. Small cells appeared to be stimulated more by grafting than by wounding. Larger coelomocytes were not stimulated by grafts or wounds and were characterized by abundant cytoplasm, a few free ribosomes, polyribosomes, moderate amounts of endoplasmic reticulum, and vesicular elements incorporated ^{3}H-thymidine more frequently than did small coelomocytes characterized by a high nucleocytoplasmic ratio and many free ribosomes. After second xenografts, however, the situation is reversed, with more smaller cells labelled than large cells. Coelomocytes do respond to mitogens, wounds, and grafts – responses that show some degree of specificity and perhaps proliferation. According to Lemmi [38] the proliferative event probably lasts during 3–4 days. Using thymidine, of high specific activity to derange the DNA in a "thymidine suicide" experiment, revealed no derangement, suggesting that increased uptake of label was due to induced coelomocyte proliferation. Coelomocytes respond in vitro by directional migration to both bacterial and tissue antigens and the responding coelomocytes (92–94%) were neutrophils (type-I granulocytes) [45].

4 Evolution of the Major Histocompatibility Complex (MHC)

4.1 General Considerations

The major histocompatibility complex (MHC) is a cluster of gene loci occupying a single chromosomal region whose products govern certain immunobiological functions [35]. For its evaluation, two other reviews are recommended [5,24]. The MHC can be divided into regions, and these, in turn, into classes. Klein uses *class* for the first time to designate regions related by their genetic origin and/or function. Three different classes are designated (Class I, Class II, and Class III) based upon differences in biochemical properties, phenotypic expression and function. For relevance to this chapter we will deal only with Class I regions.

4.2 Class I Regions

During its evolution, Schreffler et al. suggested that members of a pair of Class I regions originated from an ancestral gene by gene duplication [55]. Thus, Class I regions occur in pairs with the two members of each pair closely linked, but separated by other loci. The products of Class I regions are glycoproteins whose monomers have a mol. wt. of 45,000. They are noncovalently associated with a single chain of β_2-microglobulin. Class I antigens are expressed in the plasma membrane of all cells except some cells in terminal stages of differentiation, early embryonic cells, and some neoplasias. Mature vertebrate T- and B-lymphocytes and macrophages possess the highest concentration of Class I products, and whenever in allogeneic combinations, Class I molecules will induce the production of humoral antibodies detectable by serological methods. It is not known whether invertebrate blood cells possess Class I antigens, but there is some suggestive evidence especially with respect to β_2-microglobulin.

4.3 Histocompatibility Reactions in Invertebrates

Histocompatibility reactions in invertebrates are demonstrable in multicellular organisms, although the term cytoincompatibility can be used for unicellular protozoans [15]. Recognition of non-self has been shown to occur in sponge cells (*specific cell aggregation*) of the phylum Porifera and in embryonic tissues, recalling the original definition that Class I antigens are expressed in varying degrees on plasma membranes. Class I molecules stimulate strong effector T-cell production, such as acute allograft reactions in vivo and cell-mediated lysis (CML) in vitro – the characteristics of Class I regions studied extensively from the phylogenetic perspective. The results from this approach to the MHC strongly support the argument offered by Ohno that the gene(s) controlling any reactions evolved from an earlier gene [50]. The origin of the primordial gene can be traced through phylogeny by analyzing progressively more primitive species.

5 β_2-Microglobulin-Like Molecules in Invertebrates

β_2-Microglobulin (β_2-m), present on the surface of lymphocytes and many other body cells, is a single polypeptide chain of 99 amino acid residues first isolated from human urine [2]. The precise function of β_2-m in mammals remains unclear, although it is a protein noncovalently associated with MHC heavy chains (HLA or H-2) on the plasma membrane; probably by virtue of such association, overall conformational structure of MHC molecules might be retained intact [48]. The primary sequence of the human β_2-m shows a high degree of homology with the C_H3 domain of human IgG_1, suggesting its origin from a common ancestral gene for Ig and Class I histocompatibility antigens, involved in organ transplantation [52].

In searching for the origin of β_2-m, Shalev et al. recently have demonstrated β_2-microglobulin-like molecules in total extracts of earthworms and several other invertebrates, by using radioimmunoassays [56]. Their analysis did not involve, however, a search for β_2-m in association with cell membranes (Table 1). Recently, Roch and Cooper have demonstrated β_2-microglobulin-like molecules on leukocyte membranes of the earthworm *Lumbricus terrestris* using polyclonal and monoclonal anti-human β_2-microglobulin antibodies [54]. Only one cell type, the acidophil which constitutes – 15% of the total leukocyte fraction, was labelled (Table 2). There may be no correlation between β_2-microglobulin and histocompatibility antigens in earthworms as in mammals, where they are involved in organ transplantation.

Table 1. Evidence for a putative β_2-microglobulin-like molecule in earthworms

Reagent(s)	1. Three homologous and 8 heterologous radioimmunoassay systems[a] 2. Two polyclonal and 4 monoclonal anti-human β_2-microglobulin antisera[b]
Assay(s)	1. Competition between cold and radiolabeled β_2m[a] 2. Classical double-antibody fluorescence[b]
Comments on assay and results	1. Extracts made of earthworms *Lumbricus terrestris* were positive[a] 2. Two of 4 monoclonal antibodies bound to 15% earthworm leukocytes[b]
Implications for comparative immunology	1. Evolution of β_2m-genes, MHC-Class I genes and immunoglobulin genes 2. Function of β_2 microglobulin
Future direction	1. Identification and isolation of β_2m 2. Purification and biochemical characterization of MHC-Class I-antigen-like products 3. Amino acid sequence necessary

[a] From Shalev et al. [56]
[b] From Roch et al. [54]

Table 2. Preliminary evidence for a putative Thy-1-like molecule on earthworm (*Lumbricus terrestris*) coelomocytes (Cooper et al. [22])

Reagent(s)	Monoclonal anti-human neuroblastoma #390
Assay(s)	Immunocytochemistry with monoclonal Ab 390 anti-mouse immunoglobulin Ab, avidin-biotin-peroxidase, DAB
Comments on assay and results	1. Reactivity with surface membrane of about 36% coelomocytes 2. Intracellular components also stained. Significance not known 3. Both major coelomocyte groups reactive, but not all cells of these types
Implications for comparative immunology	1. Relationship to previous work by Shalev 2. Molecules structurally related to immunoglobulins may be ubiquitous throughout animal kingdom
Future direction	1. Verify with polyclonal anti-Thy-1 and other monoclonal Thy-1 2. Test for presence in other *Lumbricus* cell types 3. Isolation, purification, amino acid sequencing 4. Relate to molecular structure of related substances in vertebrates, other invertebrates 5. Possible role(s) for Thy-1 in invertebrates

6 Evolution of the Ig Superfamily

In rodents, Thy-1 is a single polypeptide of 111 amino acids with two intrachain disulfide bridges and three N-linked oligosaccharide structures [3,4,68]. Thy-1 was originally described as a brain and thymus antigen in mice [69], but detailed analysis has revealed remarkable variations in tissue distribution between and within species, but with little information on its function. Sequence analysis and more recently genetic organization of the Thy-1 gene [47] has provided further evidence that it resembles a primordial domain showing homology with constant and variable domains of Ig. Thus, one can propose a common ancestral gene which may have served as the progenitor gene for the domains [3], the progenitor gene for β_2-microglobulin, [57] the Class I and II MHC encoded polypeptides [34,36], and members of the Ig superfamily with structures resembling an Ig-like domain [30,67].

The Thy-1 gene may be closely related in evolutionary history to the ancestral gene for Ig and MHC [68] because of its evolutionary significance and structural properties. Indeed, the likeness of Thy-1 to a primordial domain has been strenghtened by the purification and characterization of a putative Thy-1 homolog from squid brain [68]. The purified glycoprotein was similar to rodent Thy-1 in amino acid composition and one tryptic peptide showed sequence homology to rodent Thy-1 as well as Ig variable domains. Finding a Thy-1 homolog in invertebrates may establish whether this molecule is more primitive in evolution than Ig and MHC which have not as yet been found in species more primitive than vertebrates [44,49] except for the primitive Thy-1 homolog of squid, the recent work on Thy-1 homolog in tunicates [41,42] and serological evidence for its presence in earthworms [22] (Table 2).

References

1. Bailey S, Miller BJ, Cooper EL (1971) Transplantation immunity in annelids II. Adoptive transfer of the xenograft reaction. Immunology 21: 81-86
2. Berggard I, Bearn AG (1968) Isolation and properties of a low molecular weight β_2-globulin occurring in human biological fluids. J Biol Chem 243: 4095-4103
3. Campbell DG, Gagnon J, Reid KBM, Williams AF (1981) Rat brain Thy-1 glycoprotein. The amino acid sequence, disulphide bonds and an unusual hydrophobic region. Biochem J 195: 15-30
4. Cohen FE, Novotny J, Sternberg MJE, Campbell DG, Williams AF (1981) Analysis of structural similarities between brain Thy-1 antigen and immunoglobulin domains. Evidence of an evolutionary relationship and a hypothesis for its functional significance. Biochem J 195: 31-40
5. Cohen N, Collins NH (1977) Major and minor histocompatibility systems of ectothermic vertebrates. In: Götze D (ed) The major histocompatibility system in man and animals. Springer, Berlin Heidelberg New York, p 313
6. Cooper EL (1965a) Rejection of body wall xenografts exchanged between *Lumbricus terrestris* and *Eisenia foetida*. Am Zool 5: 169
7. Cooper EL (1965b) A method of tissue grafting in the earthworm, *Lumbricus terrestris*. Am Zool 5: 233
8. Cooper EL (1968) Transplantation immunity in annelids I. Rejection of xenografts exchanged between *Lumbricus terrestris* and *Eisenia foetida*. Transplantation (Baltimore) 6: 322-337

9. Cooper EL (1969a) Chronic allograft rejection in *Lumbricus terrestris*. J Exp Zool 171: 69-73
10. Cooper EL (1969b) Specific tissue graft rejection in earthworms. Science 166: 1414-1415
11. Cooper EL (1974) (ed) Contemporary topics in immunobiology, vol 4. Plenum, New York
12. Cooper EL (1976) Comparative immunology. Prentice Hall, Englewood Cliffs, New Jersey
13. Cooper EL (1977a) In: Solomon JB, Horton JD (eds) Developmental immunobiology. Elsevier, North-Holland, Amsterdam
14. Cooper EL (1977b) What is the origin and function of the immune system? Trends Biochem Sci 2: 130
15. Cooper EL (1977c) Phylogenetic aspects of transplantation. In: Masshoff JW (ed) Handbuch der Allgemeinen Pathologie, vol 1/8. Transplantation. Springer, Berlin Heidelberg New York, p 139
16. Cooper EL (1980) Cell mediated memory in invertebrates. In: Manning MJ (ed) Phylogeny of immunological memory. Elsevier/North Holland Biomedical, New York
17. Cooper EL (1985) Leukocyte activity during earthworm inflammatory responses. Tissue Reactions (in press)
18. Cooper EL, Roch P (1984) Earthworm leukocyte interactions during early stages of graft rejection. J Exp Zool 232: 67-72
19. Cooper EL, Roch P (1985) Second-set allograft responses in the earthworm *Lumbricus terrestris:* kinetics and characteristics. Transplantation (Baltimore) (in press)
20. Cooper EL, Rubilotta L (1969) Allograft rejection in *Eisenia Foetida*. Transplantation (Baltimore) 8: 220-223
21. Cooper EL, Winger LA (1975) Transplantation immunity in annelids III. Effects of temperature on xenograft rejection in earthworms. Am Zool 15: 7-11
22. Cooper EL, Stein EA, Wright RK, Klempau AE (1984) Phylogenetic distribution of a Thy-1-like determinant. Am Zool 23: 901
23. Darwin C (1900) The formation of vegetable mould through the action of worms with observations on their habits. Appleton, New York, p 326
24. DuPasquier L (1976) Phylogenesis of vertebrate immune system. In: Melchers F, Rajewski R (eds) The immune system. Springer, Berlin Heidelberg New York, p 101
25. Duprat P (1964) Mise en evidence de réactions immunitaires dans les homogreffes de paroi du corps chez le lombricien *Eisenia foetida* typica. C R Acad Sci Paris Ser C 259: 4177-4179
26. Duprat P (1967) Etude de la prise et du maintien d'un greffon de paroi du corps chez le lombricien *Eisenia foetida*. Ann Inst Pasteur 113: 867-881
27. Duprat P (1970) Specific allograft reactions in *Eisenia foetida*. Transplant Proc 2: 222-225
28. Garland JM (1978) The T-cell paradigm. A philosophical view of immunology. Dev Comp Immunol 2: 39-44
29. Gershwin ME, Cooper EL (1978) (eds) Animal models of comparative and developmental aspects of immunity and disease. Pergamon, New York
30. Hahn GS, Hamburger RN (1981) Evolutionary relationship of thymopoietin to immunoglobulins and cellular recognition molecules. J Immunol 126: 459-462
31. Hildemann WH, Benedict AA (1975) (eds) Advances in experimental medicine and biology, vol 64. Plenum, New York
32. Hostetter K, Cooper EL (1972) Coelomocytes as effector cells in earthworm immunity. Immunol Commun 1: 155-183
33. Hostetter RK, Cooper EL (1973) Cellular anamnesis in earthworms. Cell Immunol 9: 384-392
34. Kaufman JF, Strominger JL (1982) HLA-DR light chain has a polymorphic N-terminal region and a conserved immunoglobulin-like-C-terminal region. Nature 297: 694-697
35. Klein J (1977) Evolution and function of the Major Histocompatibility System: facts and speculations. In: Götze D (ed) The Major Histocompatibility System in man and animals. Springer, Berlin Heidelberg New York, p 340
36. Korman AJ, Auffray C, Schamboeck A, Strominger JL (1982) The amino acid sequence and gene organization of the heavy chain of the HLA-DR antigen: homology to immunoglobulin. Proc Natl Acad Sci USA 79: 6013-6017
37. Lemmi CAE (1982) Characteristics of primitive leukocytes equipped with receptors for xenogeneic grafts. In: Cooper EL, Brazier MAB (eds) Developmental immunology: clinical problems and aging. Academic Press, New York, pp 115-122

38. Lemmi CAE, Cooper EL (1981) Induction of coelomocyte proliferation by xenografts in the earthworm *Lumbricus terrestris*. Dev Comp Immunol Suppl 1, 5: 73-80
39. Linthicum DS, Marks DH, Stein EA, Cooper EL (1977) Graft rejection in earthworms: an electron microscopic study. Eur J Immunol 7:871-876
40. Manning MJ, Turner RJ (1976) Comparative immunobiology. Halstead Wiley
41. Mansour MH, Cooper EL (1984) Serological and partial molecular characterization of a Thy-1 homolog in tunicates. Eur J Immunol 14: 1031-1039
42. Mansour MH, DeLange R, Cooper EL (1985) Isolation, purification and amino acid composition of the tunicate hemocyte Thy-1 homolog. J Biol Chem 260: 2681-2686
43. Marchalonis JJ (1976) (ed) Comparative immunology. Blackwell Scientific, Oxford, England
44. Marchalonis JJ (1977) Immunity in evolution. Harvard University Press, Cambridge, Massachusetts
45. Marks DH, Stein EA, Cooper EL (1979) Chemotactic attraction of *Lumbricus terrestris* coelomocytes to foreign tissue. Dev Comp Immunol 3: 277-285
46. Metchinkoff E (1892) Leçons sur la pathologie comparée de l'inflammation, Masson, Paris
47. Moriuchi T, Chang H-C, Denome R, Silver J (1983) Thy-1 cDNA sequence suggests a novel regulatory mechanism. Nature 301: 80-82
48. Nakamuro K, Tanigaki N, Pressman D (1973) Multiple common properties of human β_2-microglobulin and the common portion fragment derived from HL-A antigen molecules. Proc Natl Acad Sci USA 70: 2863-2865
49. Nisonoff A, Hopper JE, Spring SB (1975) The antibody molecule. Academic Press, New York
50. Ohno S (1970) Evolution by gene duplication. Springer, Berlin Heidelberg New York
51. Parry MJ (1978) Survival of body wall autografts, allografts and xenografts in the earthworm *Eisenia foetida*. J Invertebr Pathol 31: 383-388
52. Peterson PA, Cunningham BA, Berggard I, Edelman GM (1972) β_2-Microglobulin. A free immunoglobulin domain. Proc Natl Acad Sci USA 69: 1697-1701
53. Roch P, Valembois P, DuPasquier L (1975) Response of earthworm leukocytes to concanavalin A and transplantation antigens. Adv Exp Med Biol 64: 45-54
54. Roch PG, Cooper EL, Eskinazi DP (1983) A β_2-microglobulin-like molecule on earthworm leukocyte membranes. Eur J Immunol 13: 1037-1042
55. Schreffler DC, David CS, Passmore HC, Klein J (1971) Genetic organization and evolution of the mouse H-2 region: a duplication model. Transplant Proc 3: 176-179
56. Shalev A, Greenberg AH, Lögdberg L, Bjorck L (1981) β_2-microglobulin-like molecules in low vertebrates and invertebrates. J Immunol 127: 1186-1191
57. Smithies O, Poulik MD (1972) Initiation of protein synthesis at an unusual position in an immunoglobulin gene. Science 175: 187-189
58. Stein EA, Cooper EL (1981) The role of opsonins in phagocytosis by coelomocytes of the earthworm *Lumbricus terrestris*. Dev Comp Immunol 5: 415-425
59. Stein EA, Wojdani A, Cooper EL (1982) Agglutinins in the earthworm *Lumbricus terrestris*: naturally occurring and induced. Dev Comp Immunol 6: 513-564
60. Valembois P (1963) Etude anatomique de l'évolution de greffons hétéroplastiques de paroi du corps chez quelques lombriciens. C R Acad Sci Paris Ser D 257: 3227-3228
61. Valembois P (1968) Libération de phosphatase acide dans les cellules musculaires d'un greffon de paroi du corps chez un lombricien. J Microsc 7: 61a
62. Valembois P (1970) Etude d'une hétérogreffe de paroi du corps chez les lombriciens. Thesis, University of Bordeaux, Talence, France
63. Valembois P (1971) Rôle des leucocytes dans l'acquisition d'une immunité antigreffe spécifique chez les lombriciens. Arch Zool Exp Gen 112: 97-103
64. Valembois P (1974) Cellular aspects of graft rejection in earthworms and some other metazoa. In: Cooper EL (ed) Contemporary topics in immunobiology, vol 4. Plenum, New York, p 121
65. Valembois P, Roch P (1977) Identification par autoradiographie des leucocytes stimulés à la suite de plaies ou de greffes chez un ver de terre. Biol Cell 28: 81-82
66. Warr GH (1978) On T-cells in invertebrates. Dev Comp Immunol 2: 555-558
67. Whilmart C, Urbain J (1976) Common origin and evolution of variable and constant regions immunoglobulins. J Immunogenet (Oxf) 3: 1-14

68. Williams AF, Gagnon J (1982) Neuronal cell Thy-1 glycoprotein: homology with immunoglobulin. Science 216: 696-703
69. Williams AF, Barclay AN, Letarte-Muirhead M, Morris RJ (1976) Rat Thy-1 antigens from thymus and brain: their tissue distribution, purification and chemical composition. Cold Spring Harbor Symp Quant Biol 41: 51-62
70. Wojdani A, Stein EA, Lemmi CA, Cooper EL (1982) Agglutinins and proteins in the earthworm *Lumbricus terrestris*, before and after injection of erythrocytes, carbohydrates and other material. Dev Comp Immunol 6: 513-624
71. Wright RK, Cooper EL (1976) Phylogeny of thymus and bone-marrow-bursa cells. North-Holland, Amsterdam

Section 3 The Defense Reactions

Chapter 12 Encapsulation in Arthropods

P. GÖTZ[1]

CONTENTS

1 Introduction

Encapsulation is the common type of defense reaction in arthropods against invading pathogens and parasites, introduced foreign bodies, or injected particulate materials. Phagocytosis predominates when the haemocoel is exposed to small numbers of foreign particles [28]. Further immune capacity of arthropods is based on the presence or neosynthesis of humoral substances, such as lysozyme, cecropins, attacins (see Chap. 6), and other compounds with antibiotic efficiency [13].

Encapsulation consists of the formation of a capsule-like envelope around foreign objects. In general, this capsule is formed by attaching blood cells. Such encapsulation by blood cells takes place in two forms: (a) as *nodule formation* against small particulate material and (b) as *capsule formation* (proper cellular encapsulation) which occurs if the foreign object reaches or exceeds the size of the blood cells.

In some insects melanotic crusts are deposited on the surface of foreign objects without visible participation of blood cells. This type of defense reaction has been called *humoral or melanotic encapsulation* [11,13,14,23].

Reviews on immunity in arthropods have been published repeatedly during the last years (e.g., [13,25,27,31]). Most information concerning arthropod immunity

1 Institut für Allgemeine Zoologie (WE4), Fachbereich Biologie (23), Freie Universität, Königin-Luise-Straße 1-3, D-1000 Berlin (West), FRG

Immunity in Invertebrates
Ed. by M. Brehélin

comes from insects; little is known about crustaceans and only a few data refer to chelicerates and myriapods. The present review has been prepared for a conference on invertebrate immunity which was attended by scientists working in this field using different methods, different nomenclature, different invertebrate species and who were guided by different conceptual ideas. It is the intention of this review to tentatively draw a generalized picture of the principle events which are characteristic of the encapsulation process in arthropods and to compare it with related phenomena in other invertebrates. The reader must be aware of the fact that the outlines presented here a simplifications and generalizations which require further critical discussion. Nevertheless, they serve the purpose of allowing a check on the amount of agreement between different investigators, to provoke controversial statements, and to encourage new experiments to discriminate between alternative concepts.

During a final summing up session, the sequence of events during cellular encapsulation (listed in Table 1) and the survey of cellular reactions in different groups of invertebrates (shown in Table 2) were discussed. The participants of the final session – after supplementing additional corrections and question marks – principally accepted the content of Table 1 as a preliminary working hypothesis and Table 2 as presentation of the state of information. I hope that further investigations and future meetings will continue to improve our understanding of encapsulation and to answer some of the many question which are still open.

The review begins with an analysis of the consecutive events during cellular encapsulation in insects. Then, a description of humoral encapsulation is given, which occurs only in certain Diptera and which is interpreted as being directly related to cellular encapsulation. This is followed by a discussion of the data available from other arthropods, mainly crustaceans, in comparison to the present information on insect encapsulation. A final chapter collates cellular defense reactions of arthropods with those found in some other groups of invertebrates (annelids and molluscs).

2 Encapsulation in Insects

2.1 Cellular Encapsulation: Nodule and Capsule Formation

Cellular encapsulation has been found in all insect groups investigated. Bacteria and fungi, nematodes and eggs or larvae of parasitic insects, such as tachinids, braconids or ichneumonids are common provocators of this defense reaction, and Latex, Dowex or Sephadex beads, araldite, cellophane, cotton threads, glass needles, etc. have often

→

Fig. 1a-c. Cellular and humoral encapsulation in insects. **a** Ultrastructure of nodule from *Galleria mellonella* after injection of fungal spores (*Mucor hiemalis*). The nodule consists of melanized masses of agglutinated spores and disintegrated granular cells surrounded by concentric layers of extremely flattened plasmatocytes. *Bar* 5 μm. **b,c** Humoral encapsulation of Sephadex G beads in haemolymph of larval *Eristalomya tenax* (Diptera): **c** 30 min after addition of beads suspended in Ringer solution; **d** 60 min after addition of beads suspended in Ringer solution with phenylthiourea (PTU). Inhibition of phenoloxidase with PTU leads to the deposition of a voluminous, but soft envelope (Fig. 2 b,c: From W. Lingg [18]). *Bar* 20 μm

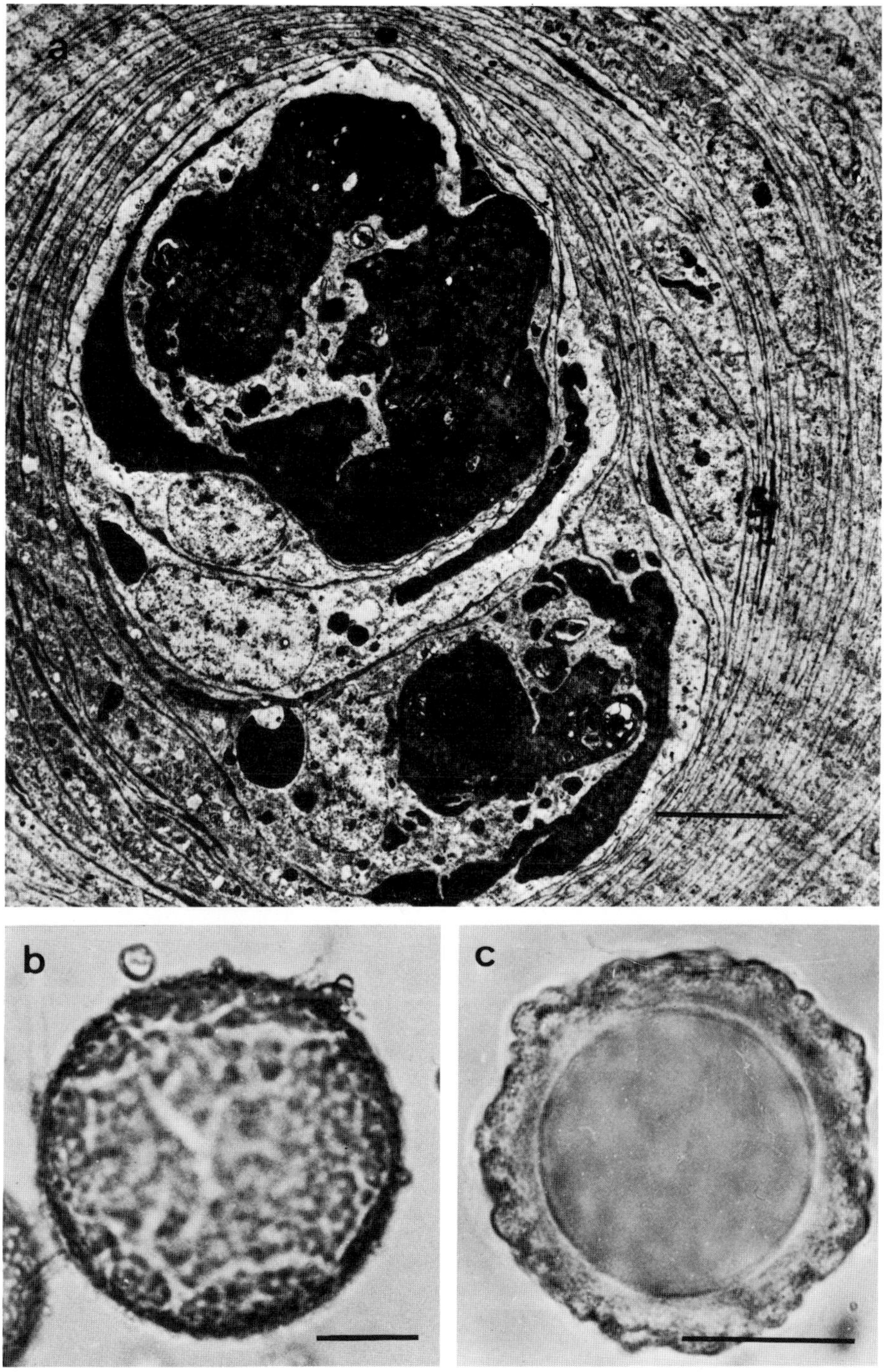
a
b
c

been used to experimentally promote the formation of a cellular envelope. In general, the reaction against living organisms is strong and always accompanied by melanization, whereas some inert materials provoke only weak reactions leading to nodules or cellular capsules with no or little melanization. Successful parasites are able to avoid cellular encapsulation, e.g. by special surface coats or by escaping into host organs (e.g. salivary gland, nerve system) where they are guarded from attacking blood cells [13,32]. Reactions to transplanted tissues depend on the generic relation between recipient and donor organism [5].

Light and electron microscopic investigations of cellular encapsulation revealed a three layered composition of the envelope [13,15,24,31]. The center of the nodule or capsule consists of the provocating agent(s) together with disintegrated and often melanized blood cells, the middle layer consists of extremely flattened blood cells, and the outer layer of rather normally shaped blood cells in loose connection (Fig. 1a). There is some evidence that two types of haemocytes are always involved in the process of cellular encapsulation – nodule as well as capsule formation. Insect blood cell types and their functions have been extensively discussed in Chap. 1.4 of this book. The blood cells, which first react to the presence of foreign material are haemocytes with apparent granules in their cytoplasm. The classification of such blood cells varies from author to author and is also dependent on the insect group in question. Inclusion-containing blood cells are also known as granulocytes, cystocytes, spherule cells or adipohaemocytes. In this review I shall use the term *granular cell (GR)* for an unstable, inclusion-containing type of blood cell which discharges its granules upon contact with foreign objects. Such activated GR's also exhibit a tendency to disintegration. The second type of blood cell which has repeatedly been found as being involved in cellular encapsulation of insects is the plasmatocyte (PL). In some insect groups other names have been used (e.g. amoebocytes, phagocytes, leukocytes, macrophages,

Table 1. Scheme of visible (LM + EM) events during cellular encapsulation (nodule and capsule formation) in insects

Time[a]	Events during cellular encapsulation
1 min	1. Random contact of blood cells with foreign material
1 min	2. Granular cells (GR's) react by degranulation
1 min	3. Discharged material sticks to foreign surfaces and to blood cells
5 min	4. Attached GR's disintegrate
15 min	5. Degranulating and disintegrating GR's attract plasmatocytes (PL's)
15 min	6. Activated PL's spread over surfaces with attached material from GR's
1–12 h	7. PL's flatten and establish cell to cell contact (desmosomes and microtubules)
1–12 h	8. Electron dense material appears in intercellular spaces between flattened PL's
6 h	9. Melanization occurs, starting from disintegrated GR's and discharged material near the foreign surface
1–3 days	10. Outermost PL's leave the capsule. Remaining capsule eventually becomes partly or totally melanized

[a] Time = approximate time after invasion or introduction of foreign objects

etc.) for cells which might represent the same functional haemocytic type. I will summarize these blood cell types under the general expression *plasmatocytes (PL's)* which are characterized as follows: PL's are rather stable blood cells with few or no granules. In contact with glass they express a tendency to spread and to form lobo- to filipode -like processes. During the second phase of encapsulation, PL's are attracted and form layers of flattened cells around the aggregates of GR's and foreign objects.

The process of cellular encapsulation has been tentatively divided into ten consecutive steps (Table 1) which shall now be discussed in detail.

2.1.1 Random Contact of Blood Cells with Foreign Surfaces

Random contact of blood cells with foreign objects occurs quite easily because of the turbulence within the blood stream. Many insects also have narrow passages within their circulatory system through which blood cells and foreign particles travel, thus, additionally increasing the opportunity for random contact.

Chemotaxis of insect haemocytes to foreign objects is discussed; however, such active movement along a gradient of chemical signals emerging from a foreign object is difficult to prove. Vey et al. [42] observed unidirectional locomotion of blood cells to fungal spores, but from their observations it is not possible to determine whether this represents chemotaxis to foreign bodies or chemotaxis to other blood cells. For example, GR's may have already come into contact with the foreign surfaces and released cell attractants.

2.1.2 Granular Cells (GR's) React by Degranulation

Circulating blood cells continuously make contact with each other and with surfaces of different tissues neighbouring the haemocoel, such as hypodermis, gut wall, muscles, fat body, nerve cords, Malpighian tubules or gonads. Under normal conditions blood cells do not react with each other and show no aggressiveness upon contact with other tissues of their own body. Exceptions are certain lethal blood cell mutants known from *Drosophila melanogaster* [9].

On contact to foreign material blood cells react by phagocytosis or encapsulation. The reason for the different behaviour of blood cells to self and non-self has to be sought in different surface qualities. All body cells of a specific animal can be expected to have identical or related surface antigens. Furthermore, blood cells and tissues which are surrounded by haemolymph may be coated with the same haemolymph molecules, and finally, the organs bordering the haemocoel may all be lined with identical basement lamellae and/or sheaths of connective tissue. In this connection it is interesting to see that in case of injury, haemocytic reactions occur only against the wounded and newly exposed area of an tissue.

According to Gagen and Ratcliffe [8], Ratcliffe and Gagen [26], Schmit and Ratcliffe [33,34], Vey et al. [42] and others, the blood cells reacting first to non-self are GR's. Other authors give different interpretations of the blood cell type which initiates cellular encapsulation. Poinar et al. [24], for example, report attachment of PL's to the foreign surface (a parasitic nematode) as the first visible reaction. Nappi [21], when describing the encapsulation of an hymenopterous parasitoid in *Musca domestica* and *Orthellia caesarion*, noted that oenocytoids initiate encapsulation. It would be important to reinvestigate these cases, since we meanwhile know that the first reaction between blood cells and foreign bodies occurs immediately after entrance of the foreign material into the haemocoel; a preceeding reaction of GR's could therefore have been missed in these earlier investigations. The nomenclature of blood cell types may also have changed with time (see Chap. 4).

Recognition. It is obvious that the reactions which lead to the activation of blood cells must occur at the cell membrane. Cell-bound receptor molecules, cell-bound agglutinins, components of the phenoloxidase system and also differences between the electrical charge of the cell membrane and foreign surface are considered as possible factors for recognition and resulting activation [13,17,30,39]. So far there are no reliable data available concerning the events which trigger cellular defense reactions. Evidence for a possible role of the phenoloxidase system and agglutinins in foreign body recognition is presented in other Chapters (2.4, 3.4) of this book. These factors will be discussed again later, in the context of melanization and humoral encapsulation in insects.

Degranulation. A fibrillar substructure within the vacuoles becomes visible during the process of degranulation. After complete degranulation the originally electron dense vacuoles are light. The material released from GR's is electron dense (refracting under phase contrast conditions); it swells after liberation and forms halos of granular appearance around the discharging cell. It foreign objects are present in the neighbourhood of discharging GR's, the material reveals its adhering nature and sticks to the foreign surfaces. Often GR's not only discharge the contents of their vacuoles, but disintegrate completely, leaving cell debris spread over an area much larger than the original cell diameter. Degranulation of inclusion-containing blood cells may represent a diversity of chemical and biological processes. GR's seem to have more than one function or consist of several subclasses of blood cells, each with different capabilities. Activated GR's are thought to release such different compounds as coagulogens, agglutinins, lysozyme, sticky proteins (components of the phenoloxidase system?) and also chemotactic factors attracting plasmatocytes.

2.1.3 Discharged Material Sticks to Foreign Surfaces and to Blood Cells

The following discussion is based on the assumption that GR's are in fact the first cells which react to foreign bodies by discharge of their vacuoles. The material released, possesses sticky properties and attaches preferably to foreign surfaces.

One can therefore suppose that the discharged material represents activated phenoloxidase, which is known to be sticky. This conclusion is supported by the fact that melanization occurs first in the innermost area of nodules and capsules where the contents of the granular vacuoles have attached. Occurrence of prophenoloxidase in GR's has been recently demonstrated by Ratcliffe et al. [19] after separation of insect haemocytes by gradient centrifugation (see Chap. 1).

Due to the sticky property of the discharged material, clumps of GR's and particulate material are formed or – if larger foreign objects are involved – attachment of GR's to the foreign surface occurs.

2.1.4 Attached GR's Disintegrate

Living organisms and certain nonliving materials promote heavy encapsulation which finally leads to large nodules or capsules with extended melanization. Other materials provoke a much weaker reaction with no or only little melanization. It would be useful to clarify whether the intensity of encapsulation is correlated with the proportion of granular cells which have disintegrated.

2.1.5 Degranulating and Disintegrating GR's Activate and Attract PL's. Activated PL's Spread Over Surfaces with Attached Material from GR's

The crucial point of the hypothesis presented here is the activation sequence of GR's and PL's. With regard to this aspect, there is no essential difference between nodule and capsule formation. In both cases, GR's attach to agglutinated foreign particles or to the surface of larger objects. The presence of activated GR's attracts PL's which in turn form an envelope of flattened blood cells.

It is conceivable that discharged material from granular cells induces different types of blood cells to spread and this could explain why varying blood cell types have been reported in participating in the formation of the cellular envelope.

To check the validity of this hypothetical description of an encapsulation sequence, further careful investigations are necessary. Different groups are currently devoting their activity to this problem using isolated cell populations or lysates of cells to study the capabilities of different blood cell types. Clean preparation of blood cells and optimal blood cell cultures are prerequisites for such ambitious projects (see Chap. 1).

2.1.5 PL's Flatten and Establish Cell to Cell Contact (Desmosomes and Microtubules)

Multicellular envelopes formed by flattened blood cells are typical for nodules and capsules. The expanded cells, lying in layers one above another, are very flat–sometimes measuring not more than 200 nm in thickness (Fig. 2a). The flattening of the cells is accompanied by the formation of desmosomes, microtubules and microfilaments; structures which probably contribute to the mechanical strength of such envelopes. The cells which undergo this drastic morphological change were not consistently classified as PL's. Brehélin et al. [3] in *Locusta migratoria* and *M. melolontha* recognized GR's rather than PL's as forming the middle part of the envelope. Again one would like to have these observations reaffirmed on the basis of a standard blood cell type classification system (see Chap. 4).

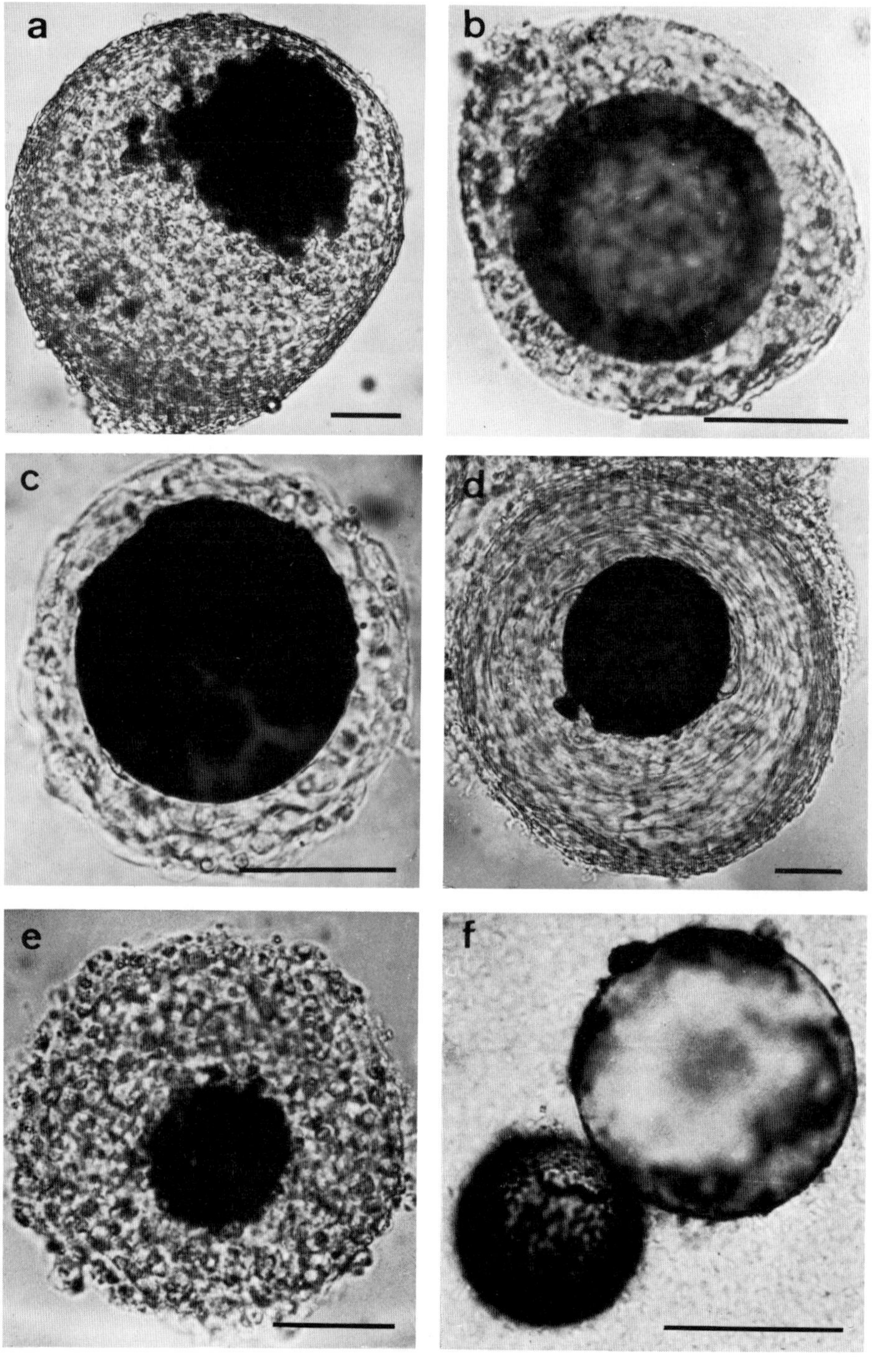
a
b
c
d
e
f

2.1.7 Electron Dense Material Appears in Intercellular Spaces Between Flattened PL's

The material in the intercellular space between the flattened cells is only visible on electron micrographs. Its chemical nature has not yet been explained. A fibrillar or collagen-like substructure of this intercellular material as reported from the brown bodies of annelids and cellular capsules in molluscs, was not observed in insects.

2.1.8 Melanization Occurs, Starting from Disintegrated GR's and Discharged Material Near the Foreign Surface

Melanization of the nodules or capsules becomes obvious only hours after the onset of encapsulation. The first area of melanization is always the surface of the foreign object or the center of the nodule with its aggregation of GR's and foreign particles. This supports the assumption that the components of the phenoloxidase cascade originate from the cells which reacted first to the foreign objects, that is the GR's.

The melanization process can continue for days, expanding continuously from the center to the middle and outer layer of the capsule. As mentioned before (2.1.4) it would be important to see whether such heavy encapsulation correlates with high proportions of disintegrated GR's.

So far, only cellular capsules around parasites (e.g. nematodes or insect parasitoids) have been found to be completely melanized. Such mummified parasites remaining in the haemocoel without further reaction, have often been found in insects as proof of a successful defense reaction [12,13,32].

2.1.9 Outermost PL's Eventually Leave the Cellular Envelope

It is necessary to explain how encapsulation eventually comes to a halt. Only the innermost blood cells had direct contact with the foreign surface. Further blood cells can only attach to bound blood cells. The concept of a biphasic type of encapsulation with GR's reacting to the foreign surface, and PL's reacting to the disintegrating GR's, offers a plausible explanation of a mechanism which finally ends the aggregation of blood cells. As soon as the GR's are covered by attaching PL's, the attracting signal which is released from the activated GR's (and perhaps from the foreign object also) becomes weaker, and with increasing thickness and tightness of the envelope, its dif-

Fig. 2a-f. Cellular and humoral encapsulation in insects. **a-d** Cellular encapsulation of inert material (Dowex 50 WX4) in larval insects from four different groups: **a** *Aeschna spec.* (Odonata); **b** *Panorpa vulgaris* (Mecoptera); **c** *Tipula spec.* (Diptera); **d** *Galleria mellonella* (Lepidoptera). The larvae have been injected with Dowex beads suspended in Ringer solution and were dissected some days later to regain the capsules. **e,f** Comparison of cellular and humoral encapsulation. Beads of Sephadex after (**e**) cellular encapsulation in larvae of *Tenebrio molitor* (Coleoptera) and after (**f**) humoral encapsulation in isolated haemolymph of larvae of *Culex pipiens* (Diptera). From W. Lingg [18] . *Bars* 50 μm

fusion may be completely inhibited. This prevents further growth of the capsule or nodule and even allows some peripheral cells to leave the aggregation.

2.2 Humoral Encapsulation in Insects

Humoral encapsulation consists of the deposition of a quickly hardening substance on the surface of foreign objects. The material is precipitated from the haemolymph without visible participation of haemocytes (Figs. 1 b,c; 2 f). Such humoral encapsulation is only known from the larvae of certain Diptera; its occurrence correlates with low haemocyte counts [15]. Humoral encapsulation acts as an effective defense reaction against nematodes, fungi and bacteria. In some species of chironomidae, invading nematodes were exclusively killed by encapsulation [10]. Among fungi tested, only the fast growing *Beauveria bassiana* overcame this defense reaction. Hyphae of *B. bassiana* provoked heavy humoral encapsulation also within the cuticle [14,43]. Certain bacteria which are highly pathogenic to other insects (e.g. *Serratia marcescens, Pseudomonas aeruginosa, Xenorhabdus nematophilus*) were not virulent to *Chironomus*. Obviously the rapidity of humoral encapsulation prevents the release of enzymes or toxins by these normally pathogenic bacteria (Götz P. et al., submitted 1985).

Humoral encapsulation occurs in vitro as well as in vivo. The reaction can be provoked by living organisms and tissues and also by certain organic compounds, such as cellulose, Sephadex, Dowex, Latex, etc. [44].

The chemical nature of the material deposited during humoral encapsulation has been assumed to be melanin because of its brown to black colouration and its resistance to solvents and enzymes [23,43]. Since inhibitors of phenoloxidase (e.g. PTU) only suppressed hardening and darkening, the complete capsule was considered to represent a protein-polyphenol complex [14]. U. Wilke (1985, in preparation) affirmed the biphasic course of humoral encapsulation; she found the "aggregation phase" (first 10–15 min) to be Ca^{2+} dependent and tyrosine independent. The following "solidification phase", which can be inhibited by PTU (Fig. 1 b,c), proved to be independent of the presence of Ca^{2+}, but was tyrosine consuming. The phenoloxidase activity of *Chironomus* haemolymph was enhanced by addition of soluble β-1,3-glucans and inhibited with *p*-NPGB and EDTA, inhibitors of different steps of the phenoloxidase cascade in crayfish and silkworm (Harmstorf, in preparation). It seems, therefore, that activation of phenoloxidase in *Chironomus* also follows the pathway proposed by Ashida and Söderhäll [2] (see also Chap. 15, this volume).

3 Encapsulation in Crustaceans and Other Invertebrate Groups

Table 2 gives a survey over available data concerning cellular defense reactions in arthropods and other groups of invertebrates (annelids and molluscs). The haemocyte types involved and the reactions performed are listed for each group.

Table 2. Comparison of cellular defense reactions in arthropods and other invertebrate groups

Taxa	Blood cell types and their reactions							
	Haemocytes involved	Phagocytosis	Degranulation	Discharged material sticky	Cell attachment	Attached cells flatten	Nodule or capsule formation	Melanization
Insecta	Granular cells	+	+	+	+ (1.)	–	+ (1.)	+
	Plasmatocytes	++	–		+ (2.)	+	+ (2.)	– (?)
Crustacea	Semigranular cells	++	+ (1.)	+	+ (1.)	(+)	+ (1.)	+
	Granular cells	+	+ (2.)	+	+ (2.)	(+)	+ (2.)	+
	Hyaline cells	++	–		(+)	(+)	–	
Diplopoda	Granular cells					+	+ (blood)	
	Plasmatocytes					+	+ (blood)	
Chelicerata	Granular cells	+						
	Hyaline cells	+						
Annelida	Granular cells	+++	–		++			"brown
	Hyaline cells	++	–			+	+ [b]	bodies"
	Chloragog cells[a]		+		+			(melanin?)
Gastropoda	Granular cells		(+)	?		?		
	Hyalinocytes	++		?		+ fibroblasts!	+	–
Bivalvia	Granular cells	++	(+)	?	+	+	+	–
	Hyalinocytes	+		?	+	+	+	–

[a] Chloragog cells contain agglutinins, opsonin, etc.
[b] Inner layer noncellular

3.1 Crustaceans

The cellular defense reactions of crustaceans closely resemble those of insects. Invasion or introduction of bacteria, fungi, parasites or inert materials (e.g. nylon threads or human hair) leads to the formation of nodules or capsules, depending on the size of the foreign objects. As in insects, these consist of three concentric layers with necrotic, melanizing blood cells in the inner part, flattened cells in the middle and normally shaped cells in the outer region of the envelope. The flattened cells of the middle layer contain bundles of microtubules and a collagen-like material filling the intercellular spaces. All three blood cell types which occur in crustaceans (semigranular cells, granular cells and hyaline cells) participate in the formation of the middle layer [4].

The mechanisms which lead to encapsulation in crustaceans were first studied in crayfish. Upon contact, crayfish blood cells attach to hyphae of the crayfish plague fungus *Aphanomyces astaci*. Degranulation of GR's is responsible for covering the hyphae with a layer of refractile material which later becomes melanized. Melanization without participation of blood cells occurs within the cuticle of the crayfish host when hyphae of the fungus are penetrating the body wall. The material deposited on the fungal cell wall was determined as activated phenoloxidase, released from GR's, which also contain the substrate for melanization. β-1,3-glucans which are present in the cell walls of algae and fungi, are capable of initiating the deposition of active phenoloxidase. By further experiments it became evident that a phenoloxidase activating system is essentially involved in cellular encapsulation of arthropods [2,39, 41].

The phenoloxidase system consists of prophenoloxidase and additional factors, one of which is a serine protease. In crustaceans the components of the phenoloxidase system are stored within semigranular and granular cells [40]. β-1,3-glucans and lipopolysaccharides from bacterial cell walls induce exocytosis of semigranular cells and activation of the discharged prophenoloxidase [37]. Activation of the phenoloxidase system can be inhibited by Ca^{2+} depletion (addition of EDTA) and by blockade of the serine protease (with p-NPGP or STI). For details see Chap. 15 of this book.

A comparable phenoloxidase system seems to exist in insects. Ashida and Dohke [1] purified a prophenoloxidase and an activating serine protease from the cuticle of *Bombyx mori*. Activation of this prophenoloxidase occurs by cleavage of a peptide of 5,000 daltons. Activated phenoloxidase of crayfish and *B. mori* sticks to many different types of foreign surfaces.

Söderhäll and Smith [40] succeeded in separating the blood populations of the shore crab, *Carcinus maenas*, and some other marine decapods by density centrifugation in Percoll gradients using EDTA/citrate buffer at low pH as an anticoagulant. Phenoloxidase activity was strictly confined to granular and semigranular cells; pure fractions of hyaline cells were free of phenoloxidase. A possible recognition mechanism based on the prophenoloxidase cascade and a discussion of its relevance for arthropod immunity are given in Chap. 15.

3.2 Myriapods and Chelicerates

There is only little information available on defense reactions in these groups of arthropods. Ravindranath [29] describes seven basic types of blood cells for Chilopoda and Diplopoda. No description of typical encapsulation and melanization in myriapods could be found, only sheath capsules have been recorded in two species of diplopods with developing acanthor larvae [29]. We know from Mercer and Nicholas [20] that larval stages of Acanthocephala in insects do not elicit typical defense reactions. The ultrastructure of such sheaths differs from that of nodules and capsules; they lack disintegrating and flattening blood cells and show no signs of melanization [11]. The parasitic larvae are able to develop within such envelopes, which may originate from the connective tissue surrounding the gut epithelium. Since myriapods have virtually been neglected by invertebrate immunologists, the lack of information on cellular encapsulation is no proof against its existence in this arthropod group. The same should be said for chelicerates, where six basic blood cell types are registered, among them again GR's and PL's, but cellular encapsulation has not been reported so far [29].

4 Cellular Defense Reactions in Annelids and Molluscs in Comparison to Arthropods

4.1 Annelids

The cells responsible for cellular defense reactions in annelids are present in the coelomic fluid. Two main cell types are involved in phagocytosis and encapsulation, hyaline and granular amoebocytes (= coelomocytes). So far, it is unclear whether the blood vascular system of annelids contains a separate population of cells or whether blood cells and coleomocytes are of identical origin. However, the defense reactions against invading or introduced foreign objects occur mainly in the coelomic cavity; therefore, the coelomocytes and their functions have to be compared with the haemocytes in the arthropod haemocoel ([7] and Chap. 11).

Annelids possess a prominent capacity for phagocytosis and encapsulation. Small foreign particles, e.g. powdered carmine, carbon or Indian ink, but also bacteria, protozoa or mammalian erythrocytes are ingested in vitro as well as in vivo. Larger objects, such as gregarines, nematodes or larvae of nematomorpha are encapsulated. Such encapsulations in annelids resemble arthropod encapsulations in that highly flattened cells form a substantial part of the cellular envelope in both cases. However, the inner layer of annelid capsules consists of a noncellular material, exhibiting, even on electron micrographs, a homogeneous transparent appearance. This layer was assumed to be of chitinous or agglutinin nature and it remains unknown as to whether it is deposited by the parasite itself, or deposited by the attached coelomocytes or from the coelomic fluid [7,23]. Results of cellular encapsulation probably remain within the annelid for a long time where they are encountered as so-called brown bodies. These were investigated by Poinar and Hess [22] who found dead nematodes, protozoans and debris of cells and tissues in the center, surrounded by a noncellular homogeneous inner layer and a cellular outer layer composed of four different types of coelomocyte.

No indications for melanization have so far been discovered in annelids. The brown colouration of capsules results from pigmented coelomocytes or enclosed eleocytes (free cells originating from the chloragog epithelium). The transparent, electron light material which forms the inner layer of newly formed capsules and of brown bodies has no similarity with the melanotic material in arthropod capsules. So far, there is no evidence that the phenoloxidase system is involved in defense reactions of annelids. This assumption is also supported by observations made during graft rejection in annelids. Xenografts in earthworms are rejected by invasion of coelomocytes which phagocytose the foreign cells until the whole implant has disappeared [7]. There are no indications of killing of the foreign cells before phagocytosis and also no events which resemble cellular encapsulation of the implanted tissue. Arthropods, however, react against foreign tissues and large parasites, by encapsulation and melanization, which isolate the implant completely from the haemocoel.

4.2 Molluscs

Two main blood cell types are noted in molluscs, namely granulocytes and hyalinocytes (see Chap. 3), which are also referred to as granular and hyaline amoebocytes or coelomocytes. The identity of names used for blood cell types in molluscs, annelids, chelicerates and crustaceans is only based on superficial similarities. No detailed biochemical or functional comparisons have been made between blood cell populations of different invertebrate groups. Both granulocytes and hyalinocytes are phagocytic, but granulocytes were found to be much more active. Granulocytes and hyalinocytes may originate from the same stem cells and represent different age classes of one cell line; with increasing age hyalinocytes transform to granulocytes [36].

Recent electron microscopic studies have shown that encapsulation of foreign bodies is carried out exclusively by granular amoebocytes [36]. In molluscs defense reaction against parasites, inserted foreign materials or implanted heterologous tissue occur predominantly within tissues (foot muscle, epidermis, digestive gland, kidney or gonads) and less frequently in the blood vessels or sinuses of the haemocoel. Immediately after introduction of a foreign object, the region where it is located becomes infiltrated by amoebocytes (see Chap. 9). The amoebocytes which are directly attached to the foreign surface flatten and extend protrusions into the irregularities of the foreign surface. The cellular envelope solidifies by interdigitation of pseudopodia. Moreover, microtubules and bundles of microfilaments appear within the flattening cells. In the middle and outer layer intercellular spaces become filled with connective tissue fibrils in an amorphous ground substance. These substances are synthesized by amoebocytes which were transformed into fibroblasts. The extent of capsule formation differs depending on the degree of compatibility of host and parasite. Thick capsules are formed in nonsusceptible hosts, whereas there is no or only little reaction against parasites in susceptible host strains. Capsules formed around biologically inert materials have only the innermost cell layers flattened and closely packed. Living autologous and homologous tissues are not encapsulated, whereas living heterologous tissue and formalin-fixed or degenerated autologous tissue provokes a severe host reaction in which the implant is completely infiltrated and encapsulated within 1 day.

It is typical for encapsulation in molluscs that the activity of the encapsulating haemocytes is not restricted to the formation of a solid capsule around the foreign organism or tissue. The haemocytes continue their activity also after formation of a cellular envelope. They infiltrate and phagocytose the foreign tissue until only a small "granuloma", mainly consisting of connective tissue, is left.

This close relation between phagocytosis and encapsulation is also demonstrated by the fact that haemocytes labelled with Indian ink (by injecting snails prior to implantation) were found to participate in encapsulation [36].

Cheng [6] interpretes encapsulation as an unsuccessful attempt at phagocytosis. The phagocytic cells flatten and attach to the surface of the foreign body which is too large to be engulfed. Evidence for chemotaxis of blood cells to certain bacteria and metazoan parasites was reported for bivalvia [6]. To study recognition and attachment in molluscs much attention has been concentrated on the possible role of agglutinins [30]. Specific sugar binding sites have been demonstrated on haemocytes of gastropods and bivalves. Attachment and phagocytosis are significantly reduced if these binding sites are blocked by adding certain sugar molecules to the haemolymph. Also, circulating factors with opsonizing and agglutinating properties are known in molluscs (see Chap. 9, this volume).

5 Conclusions

This review represents an attempt to define principles of arthropod encapsulation. Based on recent investigations it is postulated that cellular encapsulation is the result of cooperation between two different blood cell populations. Encapsulation is initiated by granular cells (GR's) which degranulate upon contact with foreign bodies and release – among others – sticky components of a phenoloxidase activating system. A second class of blood cells, plasmatocytes (PL's), are attracted by discharging GR's. They enclose the foreign objects and attached GR's with an envelope of extremely flattened haemocytes. Simultaneously, melanization takes place which seals the surface of the foreign objects and gives the nodules and capsules a permanent consistency. So far, melanization could not be demonstrated in other groups of invertebrates except arthropods. Its restriction to the phylum Arthropoda is conceivable because of the physiological characteristics of this systematic group. Arthropods possess a cuticle and a phenoloxidase system for hardening and tanning of this cuticle. This enzyme system evolved to participate in wound healing and further defense reactions. Melanin complexes furnish a solid cover on the surface of invading organisms which make further actions of haemocytes unneccessary. In other invertebrates, e.g. molluscs, cellular encapsulation is followed by progressive phagocytosis with infiltrating phagocytes devouring the enclosed foreign cells and tissues.

To date the biphasic course of encapsulation was clearly demonstrated only in some species of insects and crustaceans. There are other reports which apparently are not in agreement with the hypothesis discussed here. Figure 2 a-d demonstrates the amount of uniformity and the degree of variation encountered in cellular encapsulation of identical foreign material (Dowex 50 WX4) in larvae or different insect orders.

Are these blood cells and their reactions comparable? Their activity leads to similar results: melanized foreign bodies surrounded by multicellular envelopes; only the thickness of the capsules and the extension of cell flattening seems to vary.

New investigations are neccessary with specimens of different invertebrate groups. Reliable analysis is needed of the ultrastructure and function of the blood cells which are consecutively involved in encapsulation. This includes identification of the biochemical reactions which occur in the course of this defense reaction.

Abbreviations

GR	granular cells
p-NPGB	*p*-nitrophenyl-*p*′-guanidinobenzoate
PL	plasmatocyte
PTU	phenylthiourea
STJ	soya bean trypsin inhibitor

References

1. Ashida M, Dohke K (1980) Activation of prophenoloxidase by the activating enzyme of the silkworm *Bombyx mori*. Insect Biochem 10: 37-47
2. Ashida M, Söderhäll K (1984) The prophenoloxidase activating system in crayfish. Comp Biochem Physiol 77B: 21-26
3. Brehélin M, Hoffmann JA, Matz G, Porte A (1975) Encapsulation of implanted foreign bodies by hemocytes in *Locusta migratoria* and *Melolontha melolontha*. Cell Tissue Res 160: 283-289
4. Bauchau AG (1981) Crustaceans. In: Ratcliffe NA, Rowley AF (eds) Invertebrate blood cells. vol 2. Academic Press, New York, pp 385-420
5. Carton Y (1976) Isogenic, allogenic and xenogenic transplants in an insect species. Transplantation (Baltimore) 21: 17-22
6. Cheng TC (1981) Bivalves. In: Ratcliffe NA, Rowley AF (eds) Invertebrate blood cells, vol 2. Academic Press, New York, pp 233-300
7. Cooper EL, Stein EA (1981) Oligochaetes. In: Ratcliffe NA, Rowley AF (eds) Invertebrate blood cells, vol 2. Academic Press, New York, pp 75-140
8. Gagen SJ, Ratcliffe NA (1976) Studies on the in vivo cellular reactions and fate of injected bacteria in *Galleria mellonella* and *Pieris brassicae* larvae. J Invertebr Pathol 28: 17-24
9. Gateff E (1978) Malignant neoplasms of genetic origin in *Drosophila melanogaster*. Science 200: 1448-1459
10. Götz P (1964) Der Einfluß unterschiedlicher Befallsbedingungen auf die mermithogene Intersexualität von *Chironomus* (Dipt). Z Parasitenkd 24: 484-545
11. Götz P (1969) Die Einkapselung von Parasiten in der Hämolymphe von Chironomus-Larven (Diptera). Zool Anz, Suppl 33. Verh Zool Ges, pp 610-617
12. Götz P (1973) Immunreaktionen bei Insekten. Naturwiss Rundsch 26: 367-375
13. Götz P, Boman HG (1985) Insect immunity. In: Kerkut GA, Gilbert LJ (eds) Comprehensive insect physiology, biochemistry and pharmacology, vol 11. Pergamon, Oxford, pp 453-485
14. Götz P, Vey A (1974) Humoral encapsulation in Diptera (Insecta): defence reactions of *Chironomus* larvae against fungi. Parasitology 58: 193-205
15. Götz P, Roettgen I, Lingg W (1977) Encapsulement humoral en tant que réaction de défence chez les Diptères. Ann Parasitol Hum Comp 52: 95-97
16. Grimstone AV, Rotheram S, Salt G (1967) An electron microscope study of capsule formation by insect blood cells. J Cell Sci 2: 281-292
17. Lackie AM, Tackle G (1985) Initiation and cessation of haemocytic encapsulation in the locust *Schistocerca gregaria* and the cockroach *Periplaneta americana*. Dev Comp Immunol 9: 173-174

18. Lingg W (1976) Experimentelle Untersuchungen zur humoralen Einkapselung. Thesis, University of Freiburg, West Germany
19. Mead GP, Ratcliffe NA, Renwrantz L (1985) Insect haemocyte separation, problems and methodology. Dev Comp Immunol (in press)
20. Mercer EM, Nicholas WL (1967) The ultrastructure of the capsule of the larval stages of *Moniliformis dubius* (Acanthocephala) in the cockroach *Periplaneta americana*. Parasitology 57: 169-174
21. Nappi AJ (1973) The role of melanization in the immune reaction of larvae of *Drosophila algonquin* against *Pseudeucoila bochei*. Parasitology 66: 23-32
22. Poinar GO Jr, Hess RT (1977) Immune response in the earthworm, *Aporrectodea trapezoides* (Annelida), against *Rhabditis pellio* (Nematoda). In: Bulla LA, Cheng TC (eds) Comparative pathobiology, vol 3. Plenum Press, New York, pp 69-84
23. Poinar GO Jr, Leutenegger R (1971) Ultrastructural investigation of the melanization process in *Culex pipiens* (Culicidae) in response to a nematode. J Ultrastruct Res 36: 149-158
24. Poinar GO Jr, Leutenegger R, Götz P (1968) Ultrastructure of the formation of a melanotic capsule in *Diabrotica* (Coleoptera) in response to a parasitic nematode (Mermithidae). J Ultrastruct Res 25: 293-306
25. Ratcliffe NA (1982) Cellular defense reactions of insects. In: Frank W (ed) Immune reactions to parasites. Fischer, Stuttgart, pp 223-244 (Fortschritte der Zoologie, vol 27. Zentralbl Bakteriol Suppl 12)
26. Ratcliffe NA, Gagen SJ (1977) Studies on the in vivo cellular reactions of insects: an ultrastructural analysis of nodule formation in *Galleria mellonella*. Tissue Cell 9: 73-85
27. Ratcliffe NA, Rowley AF (1979) Role of hemocytes in defence against biological agents. In: Gupta AP (ed) Insect hemocytes. Cambridge University Press, London, pp 331-415
28. Ratcliffe NA, Walters JB (1983) Studies on the in vivo cellular reactions of insects: clearance of pathogenic and nonpathogenic bacteria in *Galleria mellonella* larvae. J Insect Physiol 29: 407-415
29. Ravindranath MH (1981) Onychophorans and myriapods. In: Ratcliffe NA, Rowley AF (eds) Invertebrate blood cells, vol 2. Academic Press, New York, pp 327-354
30. Renwrantz L, Schänke W, Harm H, Erl H, Liebsch H, Gercken J (1981) Discriminative ability and function of the immunobiological recognition system of the snail *Helix pomatia*. J Comp Physiol 141: 477-488
31. Rowley AF, Ratcliffe NA (1981) Insects. In: Ratcliffe NA, Rowley AF (eds) Invertebrate blood cells, vol 2. Academic Press, London, pp 421-488
32. Salt (1963) The defense reactions of insects to metazoan parasites. Parasitology 53: 527-642
33. Schmit AR, Ratcliffe NA (1977) The encapsulation of foreign tissue implants in *Galleria mellonella* larvae. J Insect Physiol 23: 175-184
34. Schmit AR, Ratcliffe NA (1978) The encapsulation of araldite implants and recognition of foreignness in *Clitumnus extradentatus*. J Insect Physiol 24: 511-521
35. Sherman RG (1981) Chelicerates. In: Ratcliffe NA, Rowley AF (eds) Invertebrate blood cells, vol 2. Academic Press, New York, pp 355-384
36. Sminia T (1981) Gastropods. In: Ratcliffe NA, Rowley AF (eds) Invertebrate blood cells, vol 2. Academic Press, New York, pp 191-232
37. Smith VJ, Söderhäll K (1983) Induction of degranulation and lysis of haemocytes in the freshwater crayfish, *Astacus astacus*, by components of the prophenoloxidase activating system in vitro. Cell Tissue Res 233: 295-303
38. Smith VJ, Söderhäll K (1983) β-1,3-glucan activation of crustacean hemocytes in vitro and in vivo. Biol Bull (Woods Hole) 164: 299-314
39. Söderhäll K (1982) Prophenoloxidase activating system and melanization – a recognition mechanism of arthropods? A review. Dev Comp Immunol 6: 601-611
40. Söderhäll K, Smith V (1983) Separation of the hemocyte populations of *Carcinus maenas* and other marine decapods and phenoloxidase distribution. Dev Comp Immunol 7: 229-239
41. Unestam T, Söderhäll K (1977) Soluble fragments from fungal cell walls elicit defence reactions in crayfish. Nature 267: 45-46

42. Vey A, Bouletreau M, Quiot JM, Vago C (1975) Etude in vitro en microcinématographie des réactions cellulaires d'invertébrés vis-à-vis d'agents bactériens et cryptogamiques. Entomophaga 20: 337-351
43. Vey A, Götz P (1975) Humoral encapsulation in Diptera (Insecta): Comparative studies in vitro. Parasitology 70: 77-86
44. Wilke U (1979) Humorale Infektabwehr bei *Chironomus*larven. In: Rathmayer W (ed) Verh Dtsch Zool Ges 1979, Fischer, Stuttgart, p 315

Chapter 13 Cellular Immune Responses and Their Genetic Aspects in *Drosophila*

A. J. NAPPI[1] and Y. CARTON[2]

CONTENTS

1 Introduction

Despite the absence of immunoglobulin-like recognition molecules, the immune reactive blood cells (hemocytes) of invertebrates are capable of responding with a high degree of specificity against a diversity of nonself components. Foreign objects too large to be phagocytosed are sequestered by numerous blood cells that adhere to one another and to the foreign surface to form compact, multilayered cellular capsules that become melanized. Unfortunately, little is presently understood of the mechanism of nonself recognition in invertebrates. In some species, hemagglutinating sugar-specific binding proteins (lectins), which have been shown to be synthesized by and localized in the plasma membranes of various types of invertebrate blood cells, may severe a recognitive role as antibody-like molecules or opsonins [53,54]. Other substances that appear to be likely candidates for recognition molecules in invertebrates are components of the phenoloxidase system which function in melanization reactions that are generally manifested in the immune response of these animals [38,52].

A basic understanding of how organism identify and react against nonself components requires knowledge of the origin and types of immune competent cells, and of the mechanism which regulate their production and differentiation. This critically important information is unavailable for most invertebrate species. One organism about which considerable data are available on the origin, development, and involvement of specific types of blood cells in cellular immune reactions is *Drosophila melanogaster*. A substantial portion of our understanding of the immune system of this insect comes from studies of various genetic mutants which possess abnormal hematopoietic organs, and/or form melanotic tumors within the larval hemocoel.

1 Department of Biology, Loyola University of Chicago, Chicago, Illinois 60626, USA
2 Laboratoire de Biologie et Génétique évolutives, CNRS, 91190 Gif-sur-Yvette, France

Immunity in Invertebrates
Ed. by M. Brehélin

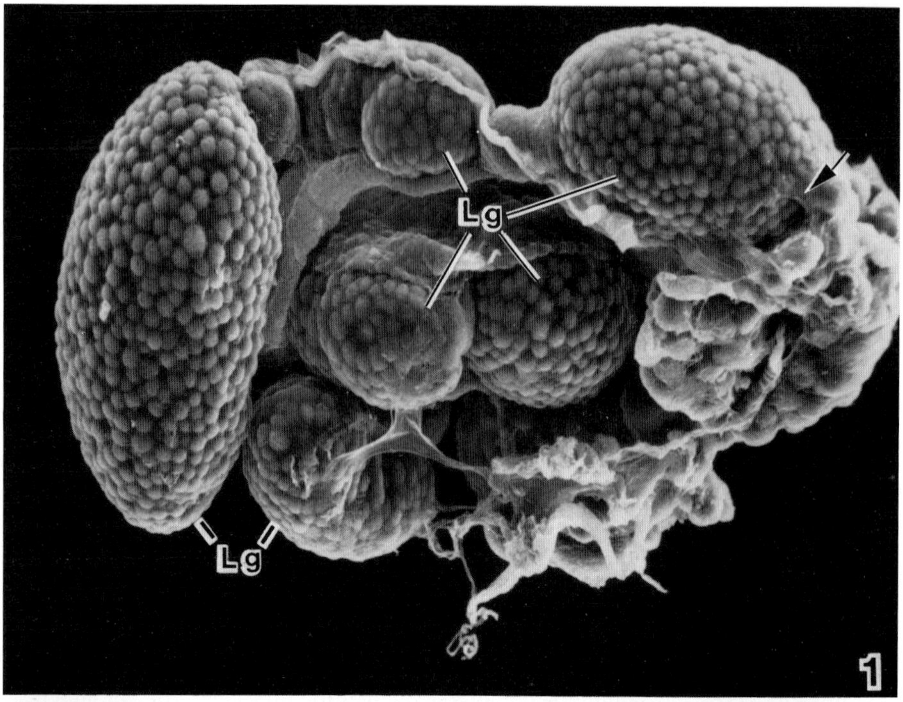

Figs. 1,2. Scanning electron micrographs of the lymph glands (*Lg*) removed from a third-instar *Basc* larva. One of the lobes (*arrow*) is enlarged in Fig. 2 and shows hemocytes within the gland and enveloped by an outer membrane (*Om*). (x 515, x 2830)

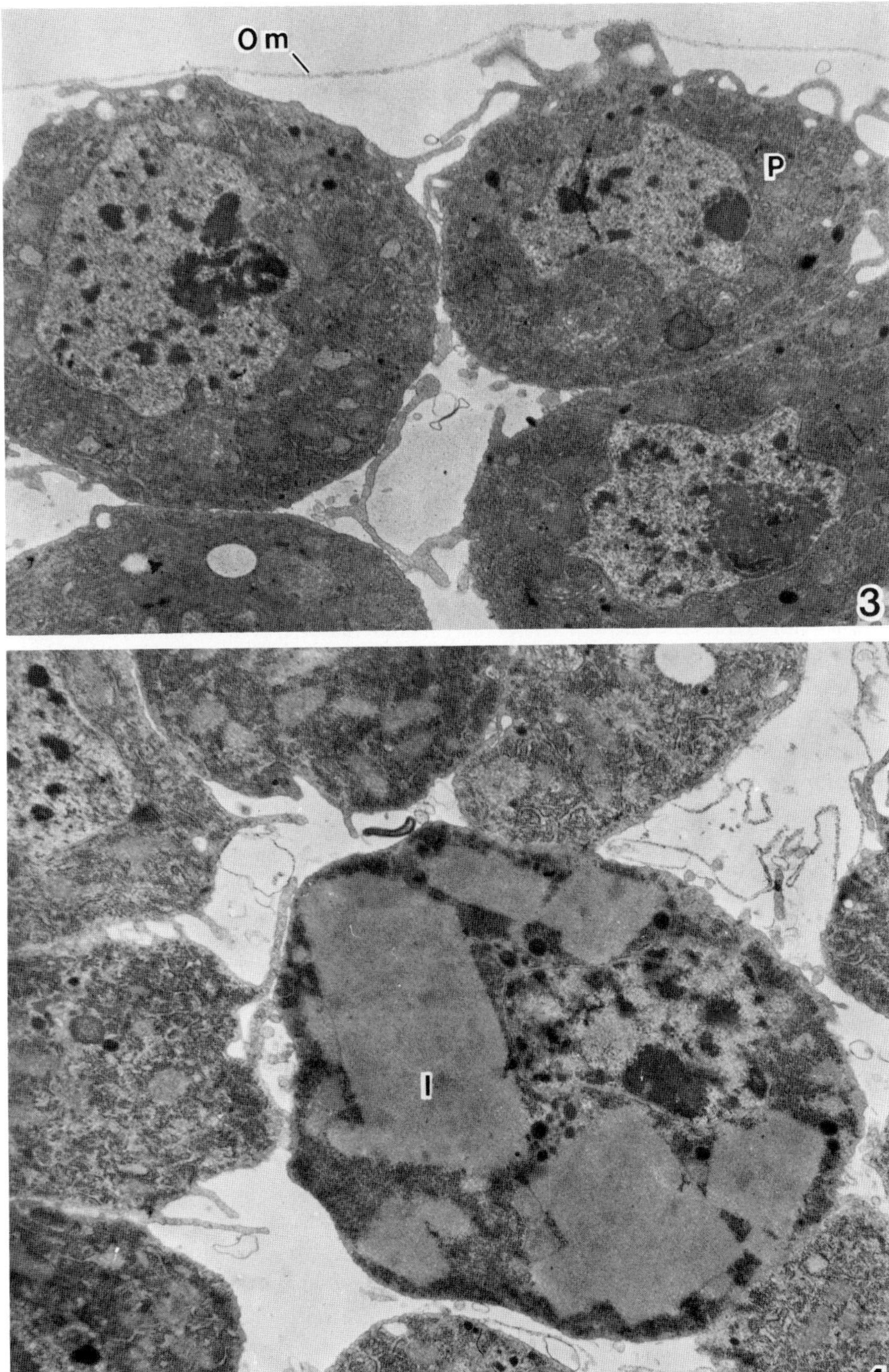

Figs. 3,4. Transmission electron micrographs of a portion of one lymph gland from a third-instar *Basc* larva. Plasmatocytes (*P*) are seen in the peripheral region of the gland just below the outer membrane (*Om*), and in a different area of the same section is seen a crystal cell containing prominent cytoplasmic inclusions (*I*). (x 8000)

2 Origin and Types of Hemocytes of *Drosophila*

The larval blood cells of *Drosophila melanogaster* are derived from five pairs of lymph glands situated along the anterior portion of the dorsal blood vessel [6,34,41,47]. Within these glands various developmental stages of two principal types of blood cell are found, the plasmatocytes and crystal cells (Figs. 1–4). It is presently uncertain whether the two types of blood cell are derived from a common stem cell (prohemocyte), or from separate progenitor cells. During larval development the hemocytes leave the hematopoietic organs and enter the circulation where they continue to differentiate, presumably under control of the endocrine system (ring glands). The plasmatocytes, which constitute the major portion of the hemocyte population throughout larval life, differentiate during development into large, flat disk-shaped cells termed lamellocytes. An intermediate cell type in the transformation of plasmatocytes to lamellocytes is the podocyte. Except for the presence of filamentous extensions of the cell surfaces of podocytes, these cells resemble plasmatocytes. The transformation of plasmatocytes to lamellocytes occurs at a high frequency toward the end of larval development [26,27]. Crystal cells, which constitute less than 10% of the hemocyte population during larval development, contain prominent paracrystalline inclusions in the cytoplasm [29]. The inclusions are typically rectangular or elongate in shape and are not membrane limited. The paracrystalline material is believed to be comprised of one or more components of the phenoloxidase complex [37].

In another *Drosophila* species (*D. yakuba*) the paracrystalline sight of crystal cell inclusions is less evident than in *D. melanogaster* [2]. Another difference is observed in lamellocytes which are more numerous in the blood of *D. yakuba* than in *D. melanogaster* (about 15% instead of 5%, 120 h old larvae under the same rearing conditions).

3 Cellular Immune Responses

The larval blood cells of *D. melanogaster* have been shown to function effectively in inflammatory and immune responses to injury, pathogens, parasites, aberrant endogenous tissues, and heterospecific tissue implants [4,12,13,15,20,22,28,30-34,36,51]. The process of phagocytosis has been reported to occur with the blood cells in circulation [20,34], as well as with the cells in the lymph glands of some mutant larvae [37]. When hemocytes are challenged in vivo and in vitro with the bacterium *Escherichia coli*, the organisms become attached to both plasmatocytes and lamellocytes. However, only the plasmatocytes have been shown to be capable of internalizing and digesting the bacterium [37]. Of interest is the finding that bacteria grown in a medium containing glucose adhere to blood cell surfaces, whereas those grown in the absence of glucose do not adhere. These studies suggest that the adhesion of bacteria to the hemocyte surface is not mediated by opsonins. There is presently no evidence that the crystal cells are involved in phagocytic activity.

During hemocyte encapsulation reactions there is a precocious transformation of large numbers of plasmatocytes to lamellocytes. These two types of hemocyte constitute the cellular component of the capsule formed around foreign objects and abnormal endogenous tissues. During capsule formation the crystal cells lyse and/or release substances that cause melanizations at the sites of these localized reactions [22, 51]. In response to infection by parasitoids, immune reactive host larvae of *Drosophila* show very high percentages of circulating lamellocytes and low percentages of crystal cells during melanotic encapsulations [3,16,20,22] (Table 1).

The precocious transformation of hemocytes which accompanies parasite encapsulation also occurs in certain mutant strains that encapsulate various larval tissues forming melanotic tumors. The involvement of hemocytes in the formation of melanotic tumors has been studied in the mutants *tu*-W, *tu*-Sz^{ts} [30-34] and *tu bw* [20]. In these mutants the melanotic growths result from hemocyte encapsulation of abnormally developing caudal adipose tissue. In the *tu*-W and *tu*-Sz^{ts} strains, encapsulation occurs in response to the degeneration of the basement membrane surrounding the adipose cells [33,34]. It is only after the basement membrane of the caudal fat body deteriorates that hemocyte encapsulation begins. In *tu bw*, however, the initial reaction of the hemocytes with the adipose tissue is cell lysis which occurs at or near intact base-

Table 1. Percentages of lamellocytes and crystal cells in larvae of *Drosophila* at various stages during parasite encapsulation reactions[a]

Hours of Parasitization	Parasitized larvae % Lamellocytes	Parasitized larvae % Crystal cells	Nonparasitized controls % Lamellocytes	Nonparasitized controls % Crystal cells
	D. melanogaster against *A. tabida* [19]			
15	3.9	0.4	1.7	2.0
20	21.0	0.1	2.4	3.3
25	33.5	0.0	2.5	5.2
30	22.4	0.0	3.2	5.7
35	11.4	0.0	3.3	6.3
	D. melanogaster against *P. mellipes* [22]			
21	15.9	1.0	0.5	2.7
33	33.2	0.3	2.7	2.2
45	29.1	1.0	7.9	2.4
57	22.0	0.6	9.4	6.1
69	15.7	1.6	3.6	5.8
104	14.8	1.8	3.9	3.1
	D. algonquin against *L. heterotoma* [16]			
27	38.2	0.0	0.0	1.5
	D. yakuba against *L. boulardi* [3]			
44 (1 egg per host)	37.0	0.94	17.7	1.24
44 (2 eggs per host)	68.1	0.44	17.7	1.24

[a] Sources of data: Nappi [19]; Nappi and Streams [22]; Nappi [16]; Carton and Kitano [3]

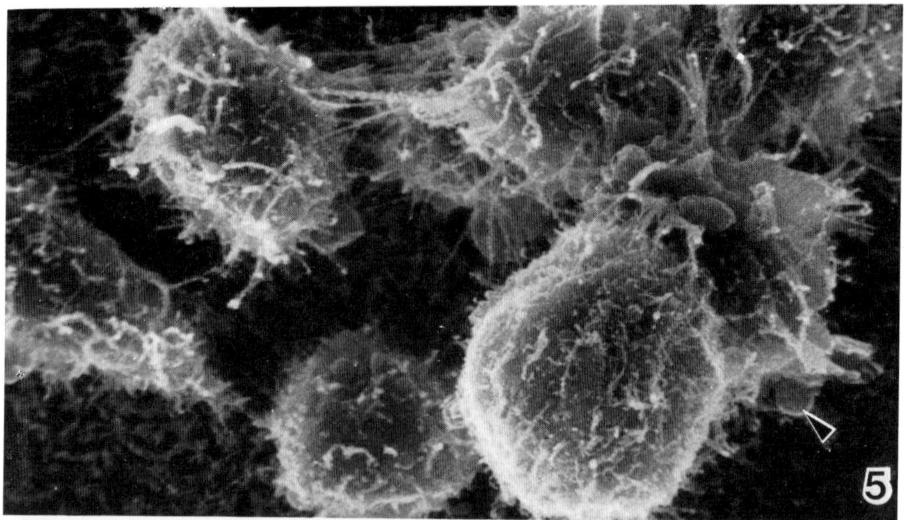
5

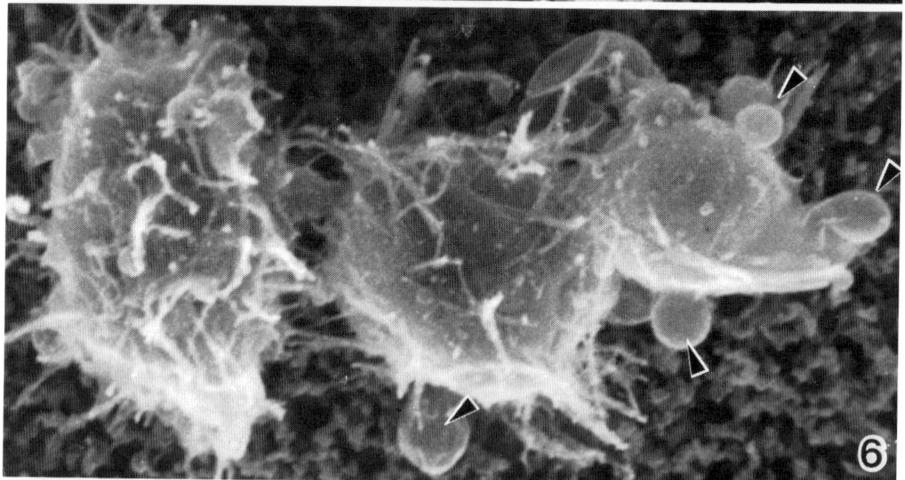
6

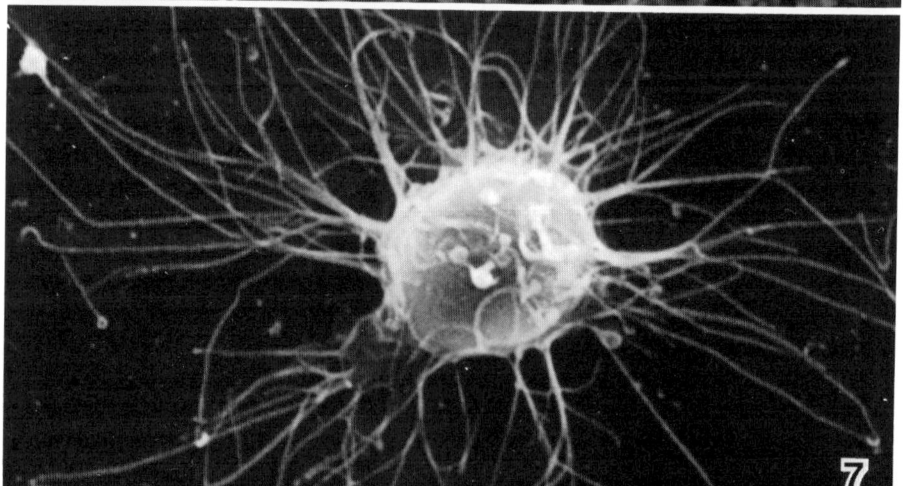
7

ment membrane. This response is rapidly followed by the aggregation of numerous hemocytes to form a multilayered, melanotic capsule. The localized blood cell reactions in this mutant coincide with aberrant, intracellular adipose cell changes. The disintegration of the basement membrane occurs only after the hemocytes have lysed at the surface and other cells have started to invade the tissue. The localized aggregation and lysis of blood cells near intact basement membrane suggests that abnormally developing adipose cells first acquire altered molecular surfaces or release substances to which the hemocytes then respond [20].

In larvae of the tumorous-lethal (Tum^1) mutant of *D. melanogaster* the lymph glands undergo neoplastic changes producing large numbers of differentiated lamellocytes [23,42]. These blood cells function abnormally in circulation by encapsulation various endogenous tissues and forming melanotic lesions [9,23]. Samples of hemolymph collected from these larvae contain large numbers of blood cells, many of which occur in small aggregations of cohering cells. When the cells are placed in culture they frequently display extensive exocytic blebbing and many prominant filamentous projections of the cell surface (Figs. 5–7). The Tum^1 mutation is temperature-sensitive for hemizygous lethality and tumorigenesis, with maximum gene expression at 29°C. Since the hematopoietic organs of Tum^1 can be triggered by a temperature shift to 29°C to produce large numbers of blood cells with the capacity to encapsulate endogenous tissues, these larvae were used to study the immune responses

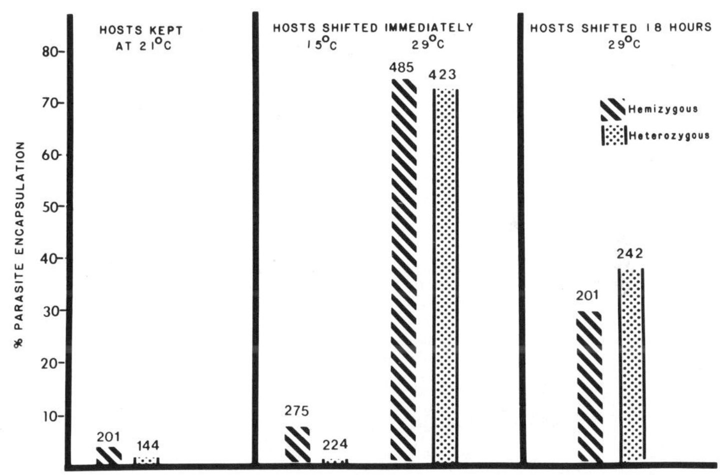

Fig. 8. Hemocyte encapsulation responses of hemizygous and heterozygous Tum^1 hosts maintained at 21°C, and shifted to 15° and 29°C following parasitization by *Leptopilina heterotoma*. Data for each temperature were pooled from tests in which hosts were infected at 52, 62, and 72 h. Data from Nappi et al. [23]

Fig. 5-7. Hemocytes of hemizygous Tum^1 larvae cultured in Shields and Sang's medium for 30 min **(Figs. 5** and **6)** and 20 h **(Fig. 7)**. Note the numerous extension of the plasma membrane between cohering cells, and evidence of globular material (*arrows*) elaborated from the cells. (x 3420, x 4760, x 3580)

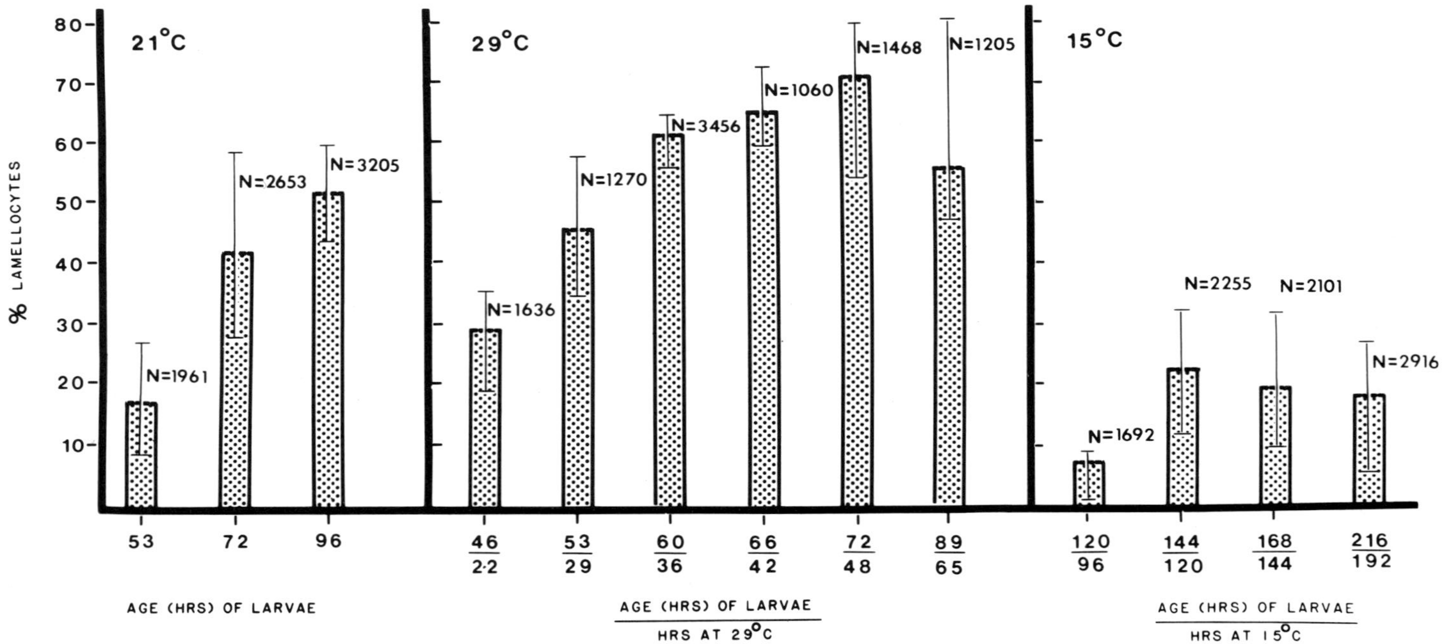

Fig. 9. Percentages of lamellocytes in hemizygous *Tum*[1] larvae reared continously at 21°C, and in larvae shifted 24 h after oviposition to 15° and 29°C. The three age invervals specified at 21°C show lamellocyte profiles for first-, second-, and third-instar larvae. At 15°C the terminal intervals are those of the first- and third-instars. At 29°C the first-instar profile is at 46 h, second at 53 to 66 h, and the third at 72 and 89 h. From Nappi et al. [23]

of the hemocytes against the parasitoid *Leptopilina heterotoma* (= *Pseudeucoila bochei*). At tumor-restrictive temperatures (15° and 21°C), parasite encapsulation is approximately 5%. However, parasite encapsulation exceeds 70% if the hosts are shifted to the tumor-permissive condition (29°C) immediately after infection. If the transfer of the hosts to the tumor-permissive condition is delayed by 18 h postinfection, parasite encapsulation drops to about 30% (Fig. 8).

Differential hemocyte counts were made to determine the lamellocyte populations in nonparasitized Tum^1 larvae, and in both immune reactive and nonreactive hosts. In nonparasitized Tum^1 reared at both 21° and 29°C there is a precocious overproduction of lamellocytes. In first-instar larvae the lamellocytes average 17% of the hemocyte populations at 21°C, and 29% at 29°C. By the middle of the third-instar these percentages increase to 51% and 70%, respectively. At 15°C, the percentage of lamellocytes range from 6.8% to 22.6% (Fig. 9). In wild-type (*Basc*) larvae lamellocytes do not appear in circulation until late in the third-instar at which time they average only about 10% of the hemocyte population [23].

The latter studies indicate that during the early stages of infection, Tum^1 larvae reared at both 21° and 29°C have sufficiently high percentages of lamellocytes in circulation to encapsulate parasites, but only at 29°C are the hemocytes immune competent. Differential blood cell counts from immune reactive and nonreactive hosts show significantly different lamellocyte profiles (Fig. 10). In immune reactive hosts that contain parasites in various stages of melanization and encapsulation at 37 h postinfection, the percentage of lamellocytes range from 43 to 65% (average, 54%). In nonreactive, susceptible hosts that contain only live parasites and show no melanization or encapsulation responses, the percentages of lamellocytes are significantly lower (average, 4%) than immune reactive hosts and nonparasitized larvae of the same age (average, 56%). In susceptible hosts the parasite suppresses the blood cell encapsulation response, presumably by producing a substance that interferes with the differentiation of plasmatocytes to lamellocytes [16,51].

The existence of this substance has been established [36] and was called lamellolysin. It was demonstrated that shortly after a female wasp oviposits in the hemocoel,

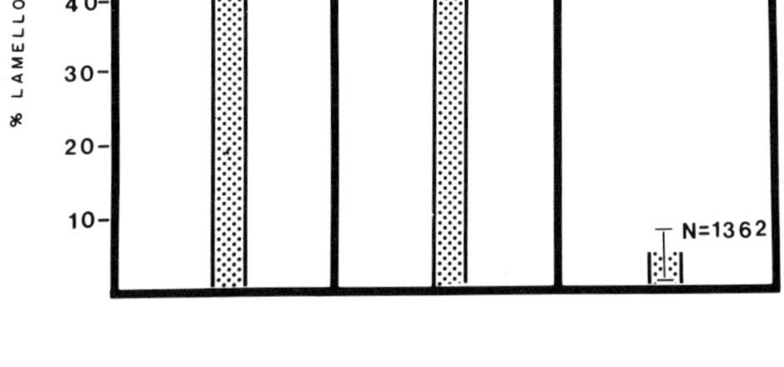

Fig. 10. Comparison of the percentages of lamellocytes in immune reactive and nonreactive hosts of the same age (89 h) as the nonparasitized controls. Blood samples from all hosts were examined 37 h after parasitization by *L. heteroma* *N*, number of cells counted from eight immune reactive hosts and ten nonreactive hosts. Data from nonparasitized larvae from Fig. 9

the lamellocytes undergo morphological changes, lose their adhesiveness and are eventually destroyed. Lamellolysin is produced by an accessory gooseneck-shaped long gland of the female reproductive system.

4 Cell Surface Alterations

The lectin-binding properties of *Tum*[1] blood cells were investigated to determine if the temperature-induced change in immune competence was associated with cell surface changes that potentiated hemocyte encapsulation of parasites and the formation of melanotic tumors. Cell surface changes were studied by treating the cells with wheat germ agglutinin (WGA) conjugated with fluorescein isothiocyanate (FITC). Although the hemocytes of control *Basc* (Muller 5) larvae exhibit no WGA-binding activity, those of *Tum*[1] larvae do bind to the lectin. At 29°C, the percentages of WGA-binding cells at each instar are more than twice as high as those at 21°C (Fig. 11). Cell counts from representative photomicrographs show that lamellocytes constitute 30% of the fluorescing cell populations at 21°C, and 60% at 29°C. When melanotic masses are collected from early-third instar larvae and treated with WGA, 20 to 30% of the cells comprising the outer layers exhibit WGA binding.

In immune reactive hosts that successfully encapsulate *L. heterotoma*, the percentages of WGA-binding hemocytes are significantly higher than in susceptible hosts that at either 21°C or 29°C, exhibit no encapsulation response (Fig. 11). The low percentages of WGA-binding hemocytes in susceptible hosts suggests that in addition to inter-

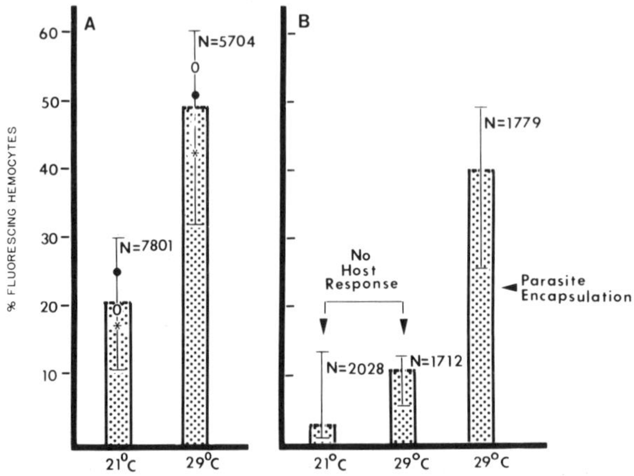

Fig. 11. A Percentages of WGA-binding hemocytes from nonparasitized hemizygous *Tum*[1] first- (*), second- (○), and third-instar larvae (•) reared continously at 21°C, and from larvae shifted 24 h after fly oviposition to 29°C. *N*, total number of cells counted from 30 larvae, ten at each of the three larval instars. Ages of larvae in hours at 21° and 29°C for each instar are (*) 52, 48; (○) 72, 60; and (•) 96,72. **B** Percentage of lectin binding hemocytes from host larvae 37 h after parasitization by *L. heterotoma*. Host age when parasitized was 52⁺ 3 h. *N*, total number of cells counted from ten larvae. *Bars* on each figure represent ranges of cell counts. From Nappi and Silvers [21]

fering with the formation of lamellocytes, the parasite also suppresses or modifies those cell surface properties that characterize the hemocytes in immune reactive hosts. Encapsulated parasites treated with WGA 24 h postinfection show approximately equal numbers of fluorescing and nonfluorescing cells in the outermost layers of the developing capsule. Encapsulated parasites examined 48 to 60 h postinfection show no WGA-binding hemocytes in the outer layer of the capsule. This suggests that either WGA-binding cells are not involved in the terminal stages of capsule formation, or that the heavy deposits of melanin obscure their WGA-binding activity.

The above studies correlate high percentages of WGA-binding hemocytes with parasite encapsulation and melanotic tumor formation in the Tum^1 mutant. It is suggested that the temperature-induced increase in immune competence of the hemocytes is due to certain cells acquiring specific surface modifications, only some of which are identified by WGA, that facilitate their adhesion. The fact that not all of the hemocytes participating in capsule formation bind to WGA suggests that there are at least two populations of immune competent blood cells; the immune competence of one population is manifested by cell surface properties that bind WGA, whereas the other population exhibits different surface features. However, it is possible that the cells lacking WGA-binding properties participate in capsule formation only by adhering to the WGA-binding cells, or to foreign surfaces coated with substances elaborated by the latter cells [35]. Studies by Rizki and Rizki [35] of the encapsulation responses of *tu*-Sz^{ts} mutants also show high percentages of WGA-binding cells at tumor-permissive conditions, and when heterospecific implants are encapsulated. They propose that WGA-binding cells extrude "xenophilic" substances that bind the cells to one another and to foreign surfaces. In the absence of WGA-binding hemocytes, the blood cells do not adhere to form capsules [37].

5 Role of Melanization

Since melanization is a common feature in both cellular and humoral immune responses of insects and other arthropods, the substrates and enzymes involved in melanin synthesis may play an important role in nonself recognition processes in these animals [7, 11,14,24,45,48,50]. Tyrosine, as well as its derivatives dopa and dopamine, are precursors of quinones which are employed as cross-linking agents in the sclerotization and melanization of the integument during preadult molting, pupariation, and adult eclosion, and in the wound healing of cuticular abrasions. The enzymes involved in these reactions (i.e., mono- and diphenoloxidases, decarboxylases, and dopamine N-acetyltransferase) may be localized intracellulary and sequestered from their substrates, and/or exist in inactive states (proenzymes). In one pathway of melanin synthesis, tyrosine is oxidized to form dopa, which is then either oxidized to form dopa quinone and ultimately melanin, or decarboxylated to form dopamine. Dopamine, a monoamine neurotransmitter with hormone-like activity, can either be oxidized to indole-5,6-quinone and then melanin, or first acetylated and then oxidized to N-acetyldopamine quinone which causes sclerotization.

Much interest has focused on the prophenoloxidase system as an important component of the immune system [44] (see Chap. 15). Recent studies have shown that enhanced phagocytosis of nonself materials in certain arthropods coincides with the activation of the prophenoloxidase system [25,43]. In the wax moth, *Galleria mellonella*, cells containing phenoloxidase (granular cells) react initially with foreign objects by discharging their contents over nonself surfaces. These coated foreign surfaces are then reacted against by other cells (plasmatocytes) which are more actively phagocytic [25]. Söderhäll et al. [46] have shown that activated phenoloxidase from crayfish hemocytes readily attaches to a variety of foreign surfaces, whereas prophenoloxidase does not attach. A serine protease is believed to be involved in both enzyme activation and attachment of the phenoloxidase to foreign surfaces. Ashida and Dohke [1] have reported similar observations for the hemolymph phenoloxidase of *Bombyx mori*.

The activation of prophenoloxidases and other enzymes involved in melanin synthesis during an immune response would result in the rapid utilization of substrates, and perhaps also an increase in the uptake of these substances by the hemocytes for use in melanotic encapsulations. The transport of melanin substrates in some insect hemocytes is believed to be mediated by a hormone-induced increase in the intracellular level of cyclic AMP, which in turn allows tyrosine to enter the cell for conversion to dopamine and other intermediary catecholamines. The subsequent release of dopamine from the hemocytes is thought to then amplify cAMP activity in other cells [49]. A dopamine-sensitive adenyl cyclase complex that modulates fluid secretion in thick salivary glands has been described by Schmidt et al. [39,40]. In addition to dopamine, two other monoamine neurotransmitters, octopamine and serotonin, have been found in nerve tissue from several insects, including *Drosophila* [10].

Recent preliminary studies suggest that some of the hemocytes of *D. melanogaster* have a dopamine-like receptor. In response to exogenous dopamine, the larval hemocytes of *D. melanogaster* exhibit in vitro cell surface changes similar to those found in immune reactive larvae. When hemocytes from 72 h old larvae of *Tum* and *Ore*-R reared at 21°C are incubated with 10^{-3} M dopamine in Shields and Sang's medium for 10 min, the percentages of WGA-binding cells are significantly higher than cells incubated without the substrate (Nappi and Silvers, unpublished). Additional studies are now warranted of the effects of dopaminergic receptor antagonists on both cAMP and cell surface alterations of the blood cells under immune reactive and nonreactive conditions.

If the phenoloxidase complex of *D. melanogaster* is involved in nonself recognition, the crystal cells would appear to play a major role in the initial reactions with foreign surfaces, since these cells are believed to be the sole source of hemolymph phenoloxidase activity [37]. The release of enzymes and melanin substrates from ruptured crystal cells may not only cause localized pigment reactions, but also induce or facilitate the differentiation of other immune competent hemocytes. Unfortunately, recent studies of hemocyte responses in the mutant Black cell (*Bc*) tend to refute this proposal. Black cell larvae lack typical crystal cells and instead have a pigmented, mutant form of these hemocytes [34]. Presumably, a gene defect causes a disruption of the integrity of the paracrystalline inclusions, resulting in phenolic substrate and enzyme interactions and the melanization of the cells. In addition, cell-free hemolymph shows no detectable phenoloxidase activity. Despite these impairments, phagocytic and encapsulation responses are not adversely affected, except that the capsules formed are amelonotic. These observations led Rizki and Rizki [37] to suggest that the crystal cells are not functional in the recognition of foreign objects. Black cell mutants appear to be a useful tool for future studies of the role of phenolic substrates and enzymes in immune responses. Since normal pigmentation of the puparium and adult cuticle occur in this mutant, cuticular phenoloxidases and decarboxylases must be functional. Unfortunately we have little information concerning the localization of these and other enzymes involved with *Drosophila* catecholamine metabolism. Because catecholamine metabolites are very reactive compounds and potentially detrimental to the insect, it seems reasonable to suggest that they are localized within epidermal cells for use in cuticular hardening and darkening, and for repairing cuticular lesions. Thus, when the cuticle is penetrated by a parasite, or ruptured during implantation of foreign objects, these compounds may become active and generate sclerotinizing and melanizing products that not only function in the repair process, but also serve in defense.

6 Genetic Aspect of Encapsulation

Encapsulation ability appears to be under partial genetic control. By comparing the "pigment reaction" level (% larvae with dispersed melanized hemocytes) and the encapsulation rate in various geographical strains of *Drosophila melanogaster* infested with the larval endoparasite *Leptopilina heterotoma* (Hymenoptera), Hadorn and Walker [8] established that these traits are not constant throughout the host species, but vary between populations. A Swiss stock appeared resistant with a strong defense reaction against parasite eggs, while an Egyptian stock showed a weak reaction.

Selection in hybrid lines between these two stocks and in Egyptian stock was effective in increasing the frequency of pigment reaction. This reaction, however, does not interfere with the development of the wasp. The ability to make a capsule appeared to have increased through selection by less than 2% per generation.

Experiments using the isofemale strain method were performed to investigate the genetic aspect of encapsulation ability [5] in *Drosophila melanogaster* (African strains). An isofemale strain is set up from natural population and each one is originated from a single female inseminated in nature. This method permits the assessment of intrapopulation genetic variation. Measurements of rate of encapsulation against a parasitic wasp (*Leptopilina boulardi*) were carried out on each of 22 isofemale strains. This rate is equal to flies number with a capsule versus attacked host number. The rate of encapsulation varied between 12% and 93% (mean = 55.7%) at the 8th generation (Fig. 12) and between 11% and 100% at the 15th generation (mean = 47.8%). The constancy of the trait on each strain was tested. The values obtained on the two different generations are highly correlated ($r = 0.62, P < 0.01$) (Fig. 13). Thus, stability with time of the mean for all the strains together is due to stability of each strain individually. By comparison with the analysis of genetic variability in the isofemale strain method, it was shown that variability between strains is significantly higher than variability within strains. We can consider variability among strains as a familial trait under genetic control.

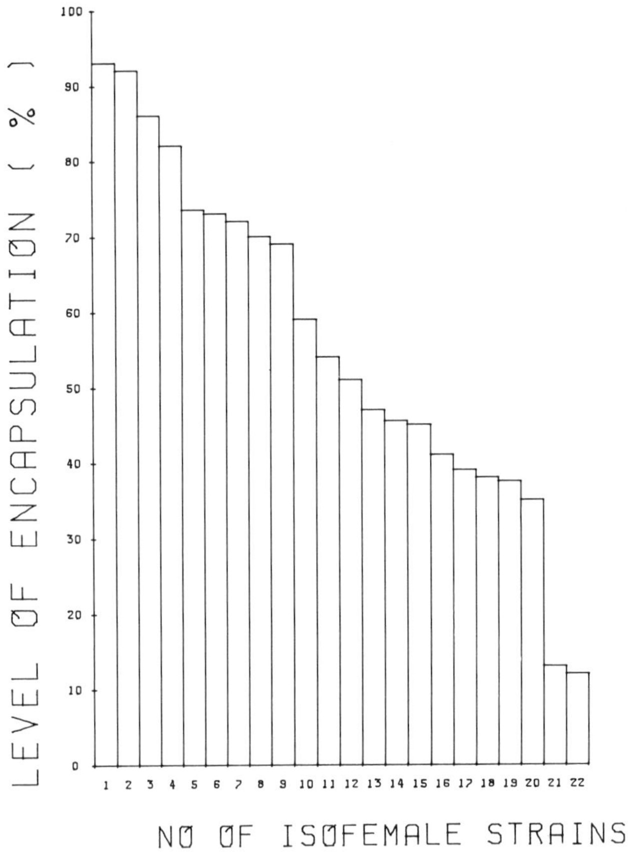

Fig. 12. Variability of the encapsulation ability of host *Drosophila melanogaster* according to the isofemale strain (the *Drosophila* strains were numbered in order of decreasing rate of encapsulation)

The genetic control of this cellular response may occur during different steps of the encapsulation process and two different possibilities at least may be envisaged to explain this mechanism. The first may be related to a genetic variability in plasmatocyte-lamellocyte transformation. Another explanation, more speculative, is to consider the possibility that the ability to set up a reaction against a specific foreign structure might be genetically regulated.

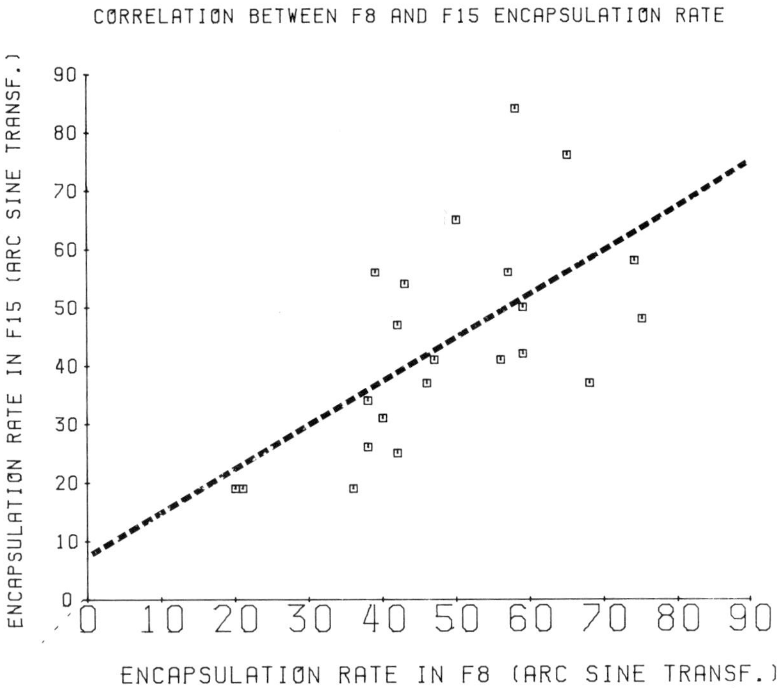

Fig. 13. Correlation between encapsulation rates of *Drosophila melanogaster* of the F_8 and F_{15} generations

References

1. Ashida M, Dohke K (1980) Activation of prophenoloxidase by the activating enzyme of the silkworm, *Bombyx mori*. Insect Biochem 10: 37
2. Brehélin M (1982) Comparative study of structure and function of blood cells from two *Drosophila* species. Cell Tissue Res 221: 607-615
3. Carton Y, Kitano H (1979) Changes in the hemocyte population of *Drosophila* larvae after single and multiple parasitization by *Cothonaspis* (Parasitic Cynipidae). J Invertebr Pathol 34: 88-89
4. Carton Y, Bouletreau M, VanLenteren JC, Van Alphen JCM (1985) The *Drosophila* parasitic wasps. In: Ashburner M, Carson HL, Thompson JN (eds) The genetics and biology of *Drosophila*, vol 3. Academic Press, New York (in press)
5. Carton Y, Bouletreau M (1985) Encapsulation ability of *Drosophila melanogaster*: a genetic analysis. Dev Comp Immunol 9 (in press)
6. Gateff E, Shrestha R, Akai H (1984) Comparative ultrastructure of wild-type and tumorous cells of *Drosophila*. In: King RC, Akai H (eds) Insect ultrastructure, vol 2. Plenum, New York pp 559-579
7. Götz P, Vey A (1974) Humoral encapsulation in Diptera (Insecta): defense reactions of *Chironomus* larvae against fungi. Parasitology 68: 193-205
8. Hadorn E, Walker I (1960) *Drosophila* und *Pseudeucoila* I. Selektionsversuche zur Steigerung der Abwehrreaktion des Wirtes gegen den Parasiten. Rev Suisse Zool 67: 216-225
9. Hanratty WP, Ryerse JS (1981) A genetic melanotic neoplasm of *Drosophila melanogaster*. Dev Biol 38: 238-249

10. Livingstone MA, Tempel BL (1983) Genetic dissection of monoamine neurotransmitter synthesis in *Drosophila*. Nature 303: 67-70
11. Messner B (1972) Die Rolle des Tyrosinasesystems in der immunologischen Abwehrreaktion bei Wirbellosen I. Insekten Zool Jahrb Physiol 76: 368-374
12. Nappi AJ (1970) Defense reactions of *Drosophila euronotus* larvae against the hymenopterous parasite *Pseudeucoily bochei*. J Invertebr Pathol 16: 408-418
13. Nappi AJ (1973a) Hemocytic changes associated with the encapsulation and melanization of some insect parasites. Exp Parasitol 33: 287-302
14. Nappi AJ (1973b) The role of melanization in the immune reaction of larvae of *Drosophila algonquin* against *Pseudeucoila bochei*. Parasitology 66: 23-32
15. Nappi AJ (1974) Insect hemocytes and the problem of host recognition of foreignness. In: Cooper EL (ed) Contemporary topics in immunobiology, vol 4. Plenum, New York, pp 207-224
16. Nappi AJ (1975a) Cellular immune reactions of larvae of *Drosophila algonquin* against the parasite *Pseudeucoila bochei*. Parasitology 70: 189-194
17. Nappi AJ (1975b) Parasite encapsulation in insects. In: Maramorosch K, Shope R (eds) Invertebrate immunity. Mechanisms of invertebrate vector-parasite relations. Academic Press, New York, pp 293-326
18. Nappi AJ (1978) Immune reactions of invertebrates to foreign materials. In: Gershwin ME, Cooper EL (eds) Animal models of comparative and developmental aspects of immunity and disease. Pergamon, New York, pp 15-24
19. Nappi AJ (1981) Cellular immune response of *Drosophila melanogaster* against *Asobara tabida*. Parasitology 83: 319-324
20. Nappi AJ (1984) Hemocyte reactions and early pathological changes during melanotic tumor formation in *Drosophila*. J Invertebr Pathol 43: 395-406
21. Nappi AJ, Silvers M (1984) Cell surface changes during parasite encapsulation and tumorigenesis in *Drosophila*. Science 225: 1166-1168
22. Nappi AJ, Streams FA (1969) Hemocytic reactions of *Drosophila melanogaster* to the parasites *Pseudeucoila mellipes* and *P. bochei*. J Insect Physiol 15: 1551-1566
23. Nappi AJ, Kmiecik J, Silvers M (1984) Cellular immune competence of a *Drosophila* mutant with neoplastic hematopoietic organs. J Invertebr Pathol 44: 220-227
24. Poinar GO Jr, Leutenegger ▪ (1971) Ultrastructural investigation of the melanization process in *Culex pipiens* (Culicidae) in response to a nematode. J Ultrastruct Res 36: 149-158
25. Ratcliffe NA, Leonard C, Rowley AF (1984) Prophenoloxidase activation: nonself recognition and cell cooperation in insect immunity. Science 226: 557-559
26. Rizki TM (1957) Alterations in the haemocyte population of *Drosophila melanogaster*. J Morphol 100: 437-458
27. Rizki TM (1962) Experimental analysis of hemocyte morphology in insects. Am Zool 2: 247-256
28. Rizki TM (1969) Hemocyte encapsulation of streptococci in *Drosophila*. J Invertebr Pathol 12: 339-343
29. Rizki TM, Rizki RM (1959) Functional significance of the crystal cells in the larvae of *Drosophila melanogaster*. J Biophys Biochem Cytol 5: 235-240
30. Rizki TM, Rizki RM (1974a) Topology of the caudal fat body of the tumor mutant of *Drosophila melanogaster*. J Invertebr Pathol 24: 37-40
31. Rizki TM, Rizki RM (1974b) Basement membrane abnormalities in melanotic tumor formation of *Drosophila*. Experientia (Basel) 30: 543-546
32. Rizki TM, Rizki RM (1978) The role of hemocytes in melanotic tumor formation. In: Bulla LA, Cheng TC (eds) Comparative pathobiologiy, vol 4. Plenum, New York, pp 85-96
33. Rizki TM, Rizki RM (1979) Cell interactions in the differentiation of a melanotic tumor in *Drosophila*. Differentiation 12: 167-178
34. Rizki TM, Rizki RM (1980) Hemocyte responses to implanted tissues in *Drosophila melanogaster* larvae. Wilhelm Roux's Arch 189: 207-313
35. Rizki TM, Rizki RM (1983) Blood cell surface changes in *Drosophila* mutants with melanotic tumors. Science 220: 73-75

36. Rizki TM, Rizki RM (1984a) Selective destruction of a host blood cell type by a parasitoid wasp. Proc Natl Acad Sci 81: 41
37. Rizki TM, Rizki RM (1984b) The cellular defense system of *Drosophila melanogaster*. In: King RC, Akai H (eds) Insect ultrastructure, vol 2. Plenum, New York, pp 579-604
38. Salt G (1970) The cellular defence reactions of insects. Cambridge University Press, London
39. Schmidt SP, Essenberg RC, Sauer JR (1981) Evidence for a D_1 dopamine receptor in the salivary glands of *Amblyomma americanum* (L). J Cyclic Nucleotide Res 7: 375-384
40. Schmidt SP, Essenberg RC, Sauer JR (1982) A dopamine sensitive adenylate cyclase in the salivary glands of *Amblyomma americanum* (L). Comp Biochem Physiol 72C: 9-14
41. Shrestha R, Gateff E (1982) Ultrastructure and cytochemistry of the cell types in the larval hematopoietic organs and hemolymph of *Drosophila*. Dev Growth Differ 24: 65-82
42. Silvers M, Hanratty WP (1984) Alterations in the production of hemocytes due to a neoplastic mutation of *Drosophila melanogaster*. J Invertebr Pathol 44: 324-328
43. Smith VJ, Söderhäll K (1983) B-1,3 Glucan activation of crustacean hemocytes in vitro and in vivo. Biol Bull (Woods Hole) 164: 299-314
44. Söderhäll K (1982) Prophenoloxidase activating system and melanization. A recognition mechanism of arthropods? A review. Dev Comp Immunol 6: 601-611
45. Söderhäll K, Unestam T (1979) Activation of serum prophenoloxidase in arthropod immunity. The specificity of cell wall glucan activation and activation by purified fungal glycoproteins of crayfish phenoloxidase. Can J Microbiol 25: 406-414
46. Söderhäll K, Häll L, Unestam T, Nyhlén L (1979) Attachment of phenoloxidase to fungal cell walls in arthropod immunity. J Invertebr Pathol 34: 285-294
47. Stark MB, Marshall AK (1931) The blood-forming organ of the larva of *Drosophila melanogaster*. Am Inst Homeopathy 23: 1204-1206
48. Taylor RL (1969) A suggested role for the polyphenol-phenoloxidase system in invertebrate immunity. J Invertebr Pathol 14: 427-428
49. Vandenberg RD, Mills RR (1974) Hormonal control of tanning by the American cockroach: cyclic AMP as a probable intermediate. J Insect Physiol 20: 623-627
50. Vey A (1979) Effect of melanization inhibitors, anti-inflammatory drugs and irradiators on insect multicellular reactions. In: Weiser (ed) Progress in invertebrate pathology. Prague, pp 229-230
51. Walker I (1959) Die Abwehrreaktion des Wirten *Drosophila melanogaster* gegen die zoophage Cynipidae *Pseudeucoila bochei* Weld. Rev Suisse Zool 66: 569-632
52. Whitcomb RF, Shapiro M, Granados RR (1974) Insect defense mechanisms against microorganisms and parasitoids. In: Rockstein M (ed) The physiology of insects. Academic Press, New York, pp 447-536
53. Yeaton R (1981) Invertebrate lectins XI. Diversity of specificity, biological synthesis and function in recognition. Dev Comp Immunol 5: 535-545
54. Yeaton R (1983) Wound responses in insects. Am Zool 23: 195-203

Chapter 14 Hemolymph Clotting in Insects

H. Bohn[1]

CONTENTS

1 Introduction

Hemolymph clotting in insects has been reviewed repeatedly over the last 15 years [20,31,32,33,35,59]. These reviews reflect the situation of clotting research in insects. The overwhelming majority of investigations was mainly concerned with the cytological aspects of clotting: which type of hemocyte cells is engaged in clotting, and how the structure of the cells is altered during the clotting process. Until recently only a few papers have been published in which a biochemical approach was attempted. During the last years, however, some progress has been made in the analysis of the biochemical events occurring during clotting. In this review emphasis is laid on this aspect

1 Zoologisches Institut der Universität, Seidlestraße 25, 8000 München 2, FRG
Supported by Deutsche Forschungsgemeinschaft

Immunity in Invertebrates
Ed. by M. Brehélin

and, therefore, only those earlier papers which are connected with this subject will be dealt with here. For other papers, the reader is referred to the reviews cited above.

Definitions

(hemolymph) plasma	hemolymph fluid after removal of the hemocytes
(hemolymph) serum	remaining hemolymph fluid after clotting
hemocyte coagulogen	clotting protein of the hemocytes
plasma coagulogen	clotting protein of the hemolymph plasma (identical with lipophorin, but in connection with clotting the expression plasma coagulogen is preferable)
coagulum	any clotted or clot-like material, irrespective of its origin, composition, or consistence
(hemocyte) gel	coagulum formed by the hemocyte coagulogen alone
(hemolymph) clot	natural coagulum formed by hemocyte and plasma coagulogen

2 Historical Outline

Hemolymph clotting was first studied under the microscope. The events taking place after the withdrawal of the hemolymph vary considerably between different species of insects. The great variety of clotting pictures has been classified and arranged into three or four different patterns [33,70]. A short outline of the most common pattern (pattern I of Grégoire) is given here as an example.

The first observation is a structural change in part of the hemocytes. The cells often show formation and retraction of vacuoles, radial expansions of cytoplasmic blebs, swelling of the nuclei, and possibly exudation of cytoplasmic and nuclear materials. Soon afterward, a cloud of granular material appears in the surroundings of these cells. Later, the granular precipitate becomes denser, finally enclosing the cell or its nuclear remnants completely within a zone of clotted material. These formations are called "islands of coagulation".

From the events observed under the microscope, a picture of blood clotting has been drawn which was largely inspired by the clotting processes known from the vertebrates. Upon some kind of stimulus a platelet-like labile cell, the coagulocyte, releases clotting activation factors into the surrounding hemolymph. The hemolymph contains a clottable protein comparable to the vertebrate fibrinogen. Under the influence of the activation factors the hemolymph clots within a more or less extended area around the cell, thus forming the island of coagulation.

The first biochemical approach toward the clarification of the clotting mechanisms was made by Bowen and Kilby [10]. Using a Tiselius apparatus for electrophoresis they compared protein patterns of serum and plasma of *Schistocerca gregaria* and found that one of the three main peaks of the plasma proteins was strongly reduced in the serum. Accordingly, the protein content of the plasma decreased from about 7.7 mg ml^{-1} to 3.5 mg ml^{-1}. The protein fraction disappearing from the serum was interpreted as the clotting protein of *Schistocerca*. In light of our recent knowledge about plasma lipoproteins this interpretation is not conclusive. Serum was obtained by dropping hemolymph into distilled water. Under these conditions a precipitation of the plasma lipoproteins is to be expected which would account for the disappearance of one fraction of proteins irrespective of whether clotting took place or not.

The participation of lipoproteins in coagulation of *Periplaneta americana* was demonstrated by Siakotos [65] through paper electrophoresis. During clotting one lipoprotein fraction disappeared completely, another was strongly reduced, while at the same time two new fractions with lower mobility appeared which were interpreted as being part of the "coagulum network".

Further progress in the analysis of the clotting processes in insects was brought about by the work of Brehélin [12,13,14] on the hemolymph coagulation of *Locusta migratoria*. In *Locusta*, during clotting, the protein content of the plasma drops to about 80%. This decline is correlated with the disappearance of a protein with low mobility as shown by PAGE and immunoelectrophoresis. The obvious clotting protein (plasma coagulogen) was isolated from the PA gels and further characterized. It turned out to be a lipo-glyco-proteic complex with a mol. wt. between 600,000 and 1,000,000. By treatment with SDS it dissociates into six different polypeptide chains (mol. wt. between 52,000 and 150,000). Brehélin already pointed out the possible double function of the lipoprotein in clotting and lipid transport. Recently, Gellissen [23] has provided evidence for the identity of the plasma coagulogen of *Locusta* with the lipophorin, the lipid transporting system of insect hemolymph.

Brehélin also made important experiments concerning the role of the hemocytes during clotting. When he just sedimented the hemocytes without removing the supernatant plasma, coagulation only occurred in a small zone immediately above the hemocyte pellet. Plasma without hemocytes did not clot at all; the clotability of the plasma was restored by the addition of hemocytes or of a hemocyte lysate. The active factor of the lysate could be sedimented by centrifugation (2 min at 2000 x g). Extracts from various tissues were unable to induce clotting of hemocyte-free hemolymph.

The second species in which clotting was studied in more detail is a cockroach, *Leucophaea maderae* [3,4,8,9]. During coagulation in *Leucophaea* a massive clot is formed. At the same time a lipoprotein disappears from the plasma (Figs. 1,2,6) and

---

Figs. 1-7. Figs. 1,2. Comparison of hemolymph plasma (*P*) and serum (*S*) from *Leucophaea* by immunoelectrophoresis with an antiserum against hemolymph proteins (*a-P*) and a specific antiserum against the plasma coagulogen (*a-Pc*). The plasma coagulogen (*arrow*) is missing in the serum

Fig. 3. The hemocyte coagulogen is present in the hemocyte gel (*G*) and hemolymph clot (*C*) in comparable concentrations

Fig. 4. After trypsin digestion of the hemolymph clot (*Cd*) plasma coagulogen is released, but none of the other plasma proteins. The precipitation line in the middle is caused by a reaction of antibodies in the unabsorbed antiserum *a-P* with antigens contained in the absorbed antiserum *a-Pc*

Figs. 5,6,7. Purification of plasma coagulogen (*Pc*) by the association/dissociation method with hemocyte gel. The dissociated plasma coagulogen (*arrow*) is almost pure as seen by immunoelectrophoresis (**Fig. 5**) and PAGE (**Fig. 6**). SDS-PAGE (**Fig. 7**) shows that the plasma coagulogen is composed of two subunits with mol. wts. of 320,000 (at the *left*) and 76,000 (at the *right*)

Abbreviations: *P* hemolymph plasma; *S* hemolymph serum; *G* hemocyte gel, after detergent wash and ultrasonication; *C* hemolymph clot, after detergent wash and ultrasonication; *Cd* trypsin-digested hemolymph clot; *a-P* antiserum against plasma proteins; *a-Pc* antiserum against plasma coagulogen; *a-G* antiserum against hemocyte coagulogen. Ref.: Figs. 1,2 [9], 5,6 [8], 7 [3]

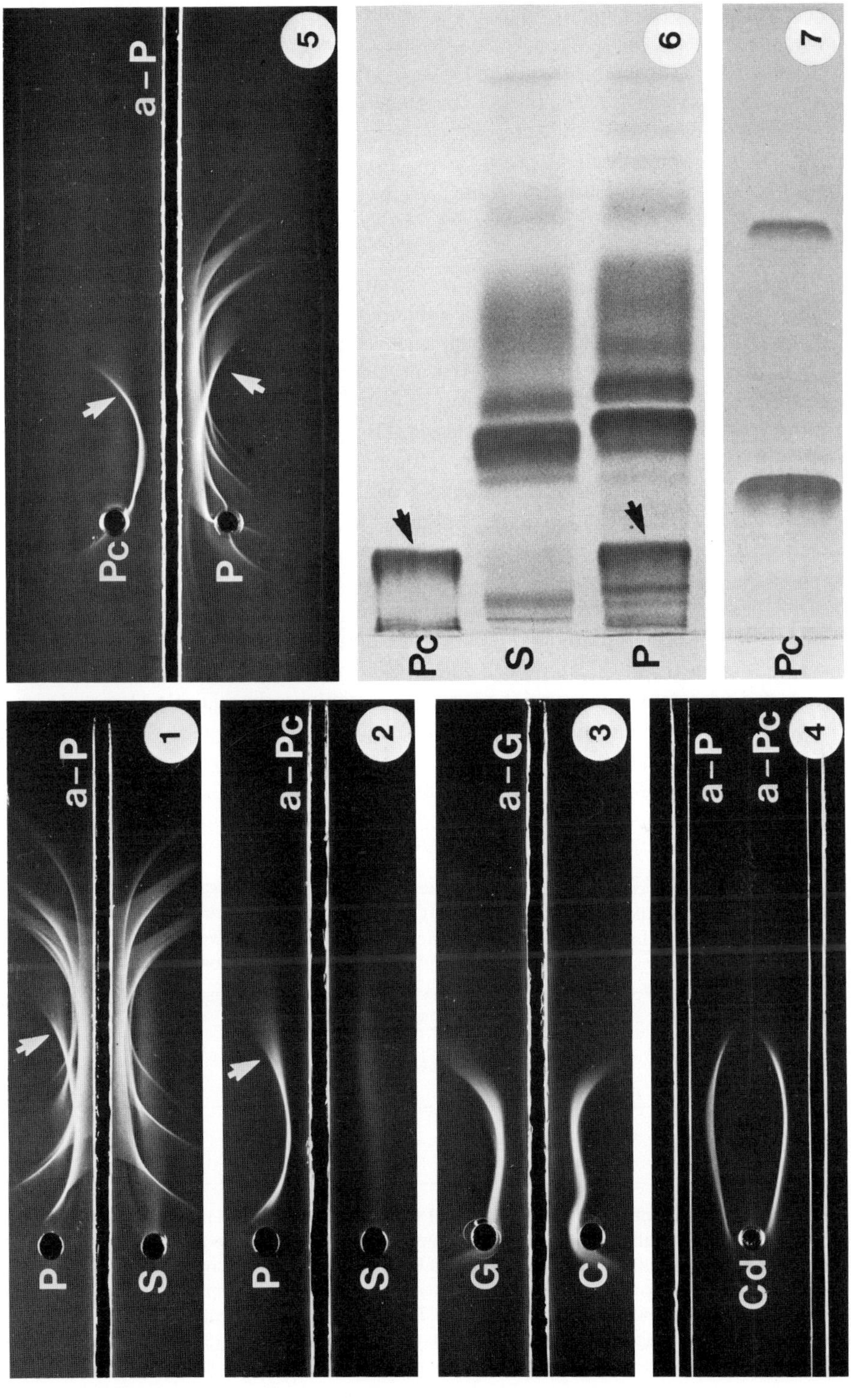
1
P
S
a–P
2
P
S
a–Pc
3
G
C
a–G
4
Cd
a–P
a–Pc
5
Pc
P
a–P
6
Pc
S
P
7
Pc

is incorporated into the clot where it can be detected in situ by immunofluorescence or by immunoelectrophoresis after homogenization or trypsin digestion (Fig. 4) of the clot. The lipoprotein, the *plasma coagulogen*, is the only plasma protein which can be found in the clot. The plasma coagulogen was purified and characterized (Figs. 5–7). It is a lipoprotein with a mol. wt. of about 640,000 and a lipid content of about 50%. The apoprotein consists of two polypeptide chains with mol. wts. of 320,000 and 76,000. The high similarity of the plasma coagulogen with the lipophorin isolated from the hemolymph of the American cockroach [18] argues for the identity of both kinds of molecules.

The importance of the hemocytes was also studied in *Leucophaea*. Plasma without hemocytes does not clot. When, on the other hand, isolated hemocytes were induced to disintegrate by freezing and thawing, surprisingly, a clot-like coagulum, which we called "hemocyte gel", was formed. The hemocyte gel does not contain any of the plasma proteins. But the gel antigens can be found in comparable concentrations in a normal clot formed in the presence of plasma (Fig. 3). This indicates that the clotting material not only consists of plasma coagulogen, but also of material derived from the hemocytes, the so-called *hemocyte coagulogen.*

In order to get an idea of how much material each of the two coagulogens contributes to the clot, equal amounts of hemocytes were induced to coagulate either in the presence or absence of plasma (Table 1). As a result, the volumes of clot and gel were nearly identical. This means that the amount of clot is mainly determined by the amount of hemocyte coagulogen. The role of the plasma in clot formation becomes evident when clot and gel are treated with 8 M urea + reducing agent (DTT). The clot resists this treatment, but the gel is nearly completely dissolved (Table 1).

According to these results the insects are provided with a unique clotting system. There are *two different clotting* proteins, *the hemocyte coagulogen, which is contained in the hemocytes* and released during the rupture of the cells, and *the plasma co-*

Table 1. Clot and gel, comparison of amounts and resistance to DTT treatment. The amount of coagulum is basically the same in both cases, but the resistance to DTT treatment is different. The clot does not dissolve in the solution (volume increase because of some swelling in the urea), while most of the gel dissolved (volume decrease). Means of four determinations ± SD

Material	Amount of coagulum		Volume increase
	prior to DTT treatment (g)	after (ml)	($ml\ g^{-1}$)
cells + plasma (clot)	0.36 ± 0.02	0.77 ± 0.15	2.10 ± 0.40
cells + Tris (gel)	0.40 ± 0.04	0.07 ± 0.02	0.17 ± 0.05

The hemocytes of 20 animals (adjusted to 0.3 g) were mixed with 0.6 ml plasma in 0.2 M Tris-HCl buffer (pH 7,2). Coagulation was induced by freezing and thawing. The resulting coagula were sedimented (5 min at 30000 x g), washed with 0.05 M Tris (10 x volume of sediment), once more sedimented and weighed (column 1). The sediments were minced and suspended in a reducing detergent solution (0.05 M DTT, 8 M urea in 0.05 Tris; 10 x vol. of sediment). After 20 h (at 37°C) the volumes of the coagula were measured by centrifugation (30 min at 1500 x g) in Nissel tubes. The volume increase was determined by dividing values of column 2 by values of column 1

agulogen which is contained in the hemolymph plasma. The hemocyte coagulogen seems to build up the basic structure of the clot which subsequently becomes stabilized and strengthened by the incorporation of the plasma coagulogen.

In the following sections, various aspects of this system – such as the molecular characteristics and origin of coagulogens (Sect. 3), their interactions (Sect. 4), and the conditions for their activity and interactions (Sect. 5) – will be described and discussed in more detail.

3 The Clotting Proteins or Coagulogens

3.1 The Plasma Coagulogen (= Lipophorin)

3.1.1 Functions

The lipophorins of insects are multifunctional molecules engaged in absorption and transport of lipids [17,39,40,42] and in hemolymph clotting. The latter function has long been overlooked; lipophorins were mainly studied in relation to their function as transport vehicles for lipid. That transport and clotting functions are performed by one and the same kind of molecules – indicated already by an earlier paper [65] – has not been fully recognized until recently [3,8,13,23].

3.1.2 Purification

A variety of procedures for the isolation and purification of the lipophorins has been developed. The decisive step in most of the procedures [3,15,16,18,24] is based on the precipitation of the lipophorin at low ionic strength, a characteristic of all lipoproteins. Shapiro et al. [64] described a simple isolation procedure by gradient centrifugation. A completely different procedure was established for *Leucophaea* [3,8] utilizing the affinity of the two coagulogens for each other. During clotting, plasma coagulogen binds to the hemocyte coagulogen. A firm connection resulting in a true clot is, however, possible only within a short time after the release of the hemocyte coagulogen from the cells. An older hemocyte gel (aged gel) still associates with the plasma coagulogen, but the association is reversible. Association is especially effective at low ionic strength (0.05 M Tris-HCl buffer) and can be reversed by increasing the ionic strength (0.2 M Tris). The association is highly specific: among the various hemolymph proteins present in the plasma only the plasma coagulogen associates. Taking advantage of these features an isolation procedure was developed. The plasma coagulogen prepared in this way (Figs. 5,6,7) was completely identical with one prepared in the usual manner by precipitation, which strongly indicates the identity of plasma coagulogen and lipophorin.

3.1.3 Molecular Composition and Structure

The molecular data of the lipophorins so far analyzed are listed in Table 2. Though one can find some differences in part of the data, the overall similarity of the lipo-

Table 2. Comparison of molecular weights and composition of lipophorins

Species	References	Molecular weight (x 10^3) Native molecule	Subunits of apoprotein Large	Small	Carbo-hydrate (%)	Lipid (%)	Lipid fractions (%)[a] DG	CH	PL	HC
Blattaria:										
Blattella germanica	[43]	511	None	75–95	0	50				
Leucophaea maderae	[3]	640	320	76	1	50	19	0	33	
Periplaneta americana	[18]	600	250	85	1	50	15	5	43	28
Saltatoria:										
Locusta migratoria	[24]	850	200	85	1	36				
Locusta migratoria	[56]	340				31	42	7	38	
Locusta migratoria	[15]	580	250	85	4	41	33	8	36	21
Lepidoptera:										
Manduca sexta	[64]		245	78						
Manduca sexta	[55]	600	285	81	2	37	34	5	38	14
Philosamina cynthia	[16]	690	370	88		44	56	13	26	1

[a] DG = diglyceride, CH = cholesterol, PL = phospholipid, HC = hydrocarbon

phorins is striking. They are lipoproteins containing between 30 and 50% lipid and 1–4% carbohydrates. The lipid moiety consists mainly of phospholipids, hydrocarbons, diglycerides, and cholesterol. The carbohydrates, consisting mostly of mannose and glucosamine, are associated with the apoprotein [15,18]. The apoprotein is composed of two noncovalently linked subunits with mol. wts. of about 250,000 and 80,000. A similar conformity is found in the amino acid composition.

Most differences are found in the data for the molecular weight of the complete native molecule, even within one and the same species. The uncertainties are due to the high lipid content which will lead to different molecular weights depending on the method used. There is also some debate about subunit composition. A large subunit does not seem to occur in *Blattella* [43], and in *Locusta*, the large subunit may or may not be found [24,25]. The inconsistency in the appearance of the large unit has led to the assumption that the high molecular subunits are polymers (trimers) of the smaller units formed as precipitation artifacts during the isolation procedure [25]. It should be mentioned in this connection that Wheeler et al. [69], under specific conditions, have found chains of lipoprotein particles under the TEM. Other authors, however, cannot find indications for a homology of the two subunits [15,18]. In *Leucophaea* the plasma coagulogens prepared in two different ways contained both large and small subunits. Recently, Shapiro et al. [64] have shown that there is no immunological cross-reaction between the larger and the smaller subunit. They also found differences in the amino acid composition. The larger subunit contains more tryptophan, but less cysteine than the smaller subunit. Most likely the native molecule is composed of one large, one small subunit, and the lipid moiety [55,64].

In view of the similarities between the lipophorins from different insect orders, it is not astonishing to find immunological cross-reaction between the lipophorins from different cockroach species [43]. Moreover, the strength of cross-reaction is in full agreement with the assumed phylogenetic relationship. Apart from the immunological cross-reaction the plasma coagulogens also show cross-reaction in clotting, i.e., plasma coagulogen of *Periplaneta* can cooperate with the hemocyte coagulogen of *Leucophaea* and form a completely normal clot (Dantanello C., Bohn H., unpublished).

3.1.4 Synthesis

The plasma coagulogen is – like other hemlymph protein – synthetized in the fat body and from this organ released into the hemolymph [25]. Synthesis of hemolymph proteins in the hemocyte has been supposed by Akai and Sato [1], but no trace of plasma coagulogen or any of the other major hemolymph proteins can be found in the hemocytes (Fig. 9) [9].

3.2 The Hemocyte Coagulogen

3.2.1 Indications for the Occurrence of a Hemocyte Coagulogen

The idea of the existence of a clotting protein in the hemocytes of arthropods is not new. In the early phase of coagulation studies on arthropods, Loeb [47,48] had already stated that *Limulus* (Xiphosura) does not have a clotting protein in the plasma, but that the clotting protein of this species is contained in the hemocytes. Loeb's observations were later fully confirmed [19,44].

Loeb [46] had also developed the concept of two phases in the coagulation process of Crustacea which later was adopted by authors studying hemolymph coagulation in insects [5,52,71]. According to this concept, a "cell coagulum" is formed in the first phase of clotting; in the following second phase – initiated by activation factors from hemocytes – plasma coagulation takes place. The cell coagulum is described as consisting mainly of agglutinated hemocytes and hemocyte remnants. The existence of a hemocyte coagulogen was not taken into consideration.

The knowledge of clotting in insects was greatly extended by the work of Grégoire (for ref. see [31,32,33,35]) who studied the clotting patterns of numerous species under the phase contrast microscope. But the idea that the hemocytes only contribute activation factors for plasma clotting was dominant. Numerous signs for the presence of a hemocyte coagulogen were neglected. Grégoire's coagulation pattern II, for example, is characterized by the formation of long cytoplasmic processes or strings between the hemocytes. Finally, a more or less dense meshwork of strings is formed into which hemocytes and plasma coagulum become embedded. These strings were interpreted by Grégoire as pseudopodia which carry along cytoplasmic granules during the "extrusion". But Marschall [50] has clearly shown that the strings develop from mucopolysaccharide-containing granules extruded from the hemocytes.

The results of electron microscopic studies of clotting cells do not give a clear answer to the question of the existence of a hemocyte coagulogen. Some authors find evidence for the extrusion of granule material from the hemocytes during clotting [51,57,60,61,63,67], while others [27,28,37] deny such a possibility. However, as long as such observations are made in the presence of plasma it is nearly impossible to decide to what extent hemocytes or plasma contribute to the material precipitated outside the cell.

3.2.2 Hemocyte Coagulogen and Gel Formation in *Leucophaea*

When the hemocytes of *Leucophaea* – in the absence of plasma proteins – are induced to disintegrate, they form a massive coagulum, superficially similar to a true clot. This coagulum resists treatment with highly concentrated solutions of denaturing agents like urea and guanidinium chloride as well as detergents or mixtures of both. The resistence of the coagulum toward detergents allows a rough purification: nuclei, cellular membranes, and other cell remnants can be washed out with a solution containing 2 M urea and 2% SDS. What remains is a homogeneous gel-like structure which we call *hemocyte gel*; the substance making up the gel is the *hemocyte coagulogen*.

So far very little is known about the chemical nature of the hemocyte coagulogen and the structure of the gel. The gel is easily digested by proteolytic enzymes and dissolves almost completely upon treatment with urea + DTT which indicates that it consists mainly of protein or that it at least has a protein backbone stabilized by disulfide bridges. The granules of most hemocytes from various insects give a positive reaction with carbohydrate-specific stains (for ref. see [2,36,62]). Polysaccharides have also been shown to be present in the clot of a lepidopteran larva [53]. This would imply that the hemocyte coagulogen is either associated with polysaccharides, or, more likely, that the hemocyte coagulogen is a glycoprotein. When detergent-treated hemocyte gel – solubilized by ultrasound treatment – is tested by immunoelectrophoresis with a specific antiserum, a slightly bipartite precipitate is formed (Fig. 3). The bipartition could indicate a chemical heterogeneity, but it may also be due to heterogeneous fragmentation through ultrasonication. There is no immunological cross-reaction between hemocyte and plasma coagulogen.

The hemocyte coagulogen is contained – obviously very densely packed – in the granules of the hemocytes (Fig. 8). During the explosion-like ejection of the material it swells considerably, presumably by hydration, forming the gel [9]. Swelling of the material is not taken into consideration by Grégoire [30] when he argues that the material contained in the hemocytes is not sufficient to account for the formation of the sometimes very extended coagulation area around the hemocytes. During gel formation or soon after, new disulfide bridges – in addition to those already existing – are formed, giving the gel its final stability. The establishment of new disulfide bridges during gel formation could be the reason for the aging of the hemocyte gel (see Sects. 3.1.2 and 5.3).

3.2.3 Coagulocytes

The existence of a specialized clotting cell, the coagulocyte, has been a matter of controversy [35,59] since it was described by Grégoire and Florkin [34]. There is general agreement on the fact that those cells which react first in clotting belong to the category of granular hemocytes. The controversy concerns the question whether among the granular hemocytes there is a subpopulation of specialized fragile cells, or whether all granular hemocytes have the possibility to function as clotting cells.

The observation that after clotting not every granular hemocyte has reacted and degranulated has often been used as an argument in favor of the existence of coagulocytes. However, in a mixture of cells the conditions are not identical in all places and not all cells will get the same stimulus. Therefore, differences in reaction are to be expected. The weakest point of the coagulocyte hypothesis is the fact that a coagulocyte can only be recognized as such after it has already reacted.

In vitro observations of *Leucophaea* hemocytes [6,7] have shown that plasmatocytes and granular hemocytes easily degranulate, while the spherule cells appear to be very stabile. Degranulation of granular hemocytes, but not of plasmatocytes, is usually connected with disintegration. The ready degranulation of plasmatocytes and granulocytes would suggest that both cell types participate in clotting and contribute material to the clot.

How different the material in the granules of the different cell types is, remains to be elucidated. According to the results of the histochemical stainings and EM observations [2,36,62] considerable differences are to be expected. Nevertheless, the granules of all three cell types show staining with fluorescent antibodies against hemocyte coagulogen (Fig. 8), also including the spherule cells which presumably do not take part in clotting. In order to get more information on the role of the different cells in clotting it would be necessary to separate the different cell types, analyze the contents of their granules, and study their clotting activity.

4 Composition and Features of the Hemolymph Clot

A true hemolymph clot develops with the participation of both coagulogens, the plasma and the hemocyte coagulogen. The consistence of the clot is considerably different from that of the hemocyte gel. The clot is resistent to DTT treatment. There are also differences in the mechanical properties, which can be deduced from the swelling behavior of gel and clot [9]. The observations suggest that the lattice work in the clot is more densely and tightly knit than in the gel.

In its composition the clot differs from the gel by the incorporation of the plasma coagulogen. Both coagulogens become firmly bound within the clot and resist any solvents, including DTT. The clot can be, however, digested by proteolytic enzymes. Upon digestion with trypsin or chymotrypsin, the plasma coagulogen is released in an almost complete form (Fig. 4).

The presence of both coagulogens in the clot raises the question of a possible interaction of the two coagulogens. Two alternatives have to be considered. Both coagulogens could form their own network independently of each other. In spite of this it is to be expected that both networks are at least partly interwoven, since they would occupy basically the same spaces. The different mechanical properties of gel and clot, for example, could be explained in this way. The other possibility is that the two coagulogens cooperate in some way during clot formation; the interaction could merely by a physical or physiological support, but could also include intermolecular connections.

The experimental evidence clearly points to the second possibility, i.e., an interaction of two coagulogens. The arguments are as follows:

1. The fact that the hemocyte coagulogen has become insoluble in DTT after clot formation is not compatible with the assumption of separate coagulation networks.
2. Both coagulogens have exactly the same distribution within the islands of coagulation as shown by immunofluorescence (Figs. 10 and 11). If the two coagulogens would coagulate independently, one would expect that only the hemocyte coagulogen is precipitated in the immediate surroundings of the cell from which it is extruded. The plasma coagulum, on the other hand, could extend well beyond the zone of the hemocyte coagulum, if its coagulation is induced by a soluble factor as is often assumed.
3. The most direct evidence for an interaction of the coagulogens is given by the association experiments (see Sect. 3.1.2), which argue for a specific affinity and binding of the plasma coagulogen to the hemocyte coagulogen.
   The idea of an association of the two coagulogens has been confirmed in experiments using specific antibodies. Association is strongly inhibited when the hemocyte gel was treated with antibodies against the hemocyte coagulogen prior to the addition of plasma [9].
   Though true clotting does not take place under conditions (aged gel!) suitable for the demonstration of association, there are good reasons to assume that a similar association of the two coagulogens also occurs during normal clotting, possibly as the first step in a series of further interactions.
4. Within a specific range of concentrations there is a fairly good correlation between the amount of coagulogen offered and the amount of clotted plasma coagulogen [8]. This might indicate that the hemocyte coagulogen has only a restricted capacity for the interaction with the plasma coagulogen, a fact which also argues in favor of an association of the coagulogens.

The data presented strongly support the idea that during clot formation the two coagulogens combine and become cross-linked in some way. Arrangement and cross-

---

Figs. 8-11. Immunofluorescent localization of hemocyte coagulogen (**Figs. 8,10**) and plasma coagulogen (**Figs. 9,11**) in hemocytes (**Figs. 8,9**) and islands of coagulation (**Figs. 10,11**) of *Leucophaea*. **a** phase contrast; **b** fluorescence pictures

Fig. 8a,b. The hemocyte coagulogen is contained mainly in the granules of the hemocytes (*G* granular hemocyte, *S* spherule cell, *P* plasmatocyte)

Fig. 9a,b. Plasma coagulogen is not contained in the hemocytes

Figs. 10a,b and 11a,b. In the island of coagulation hemocyte (**Fig. 10**) and plasma coagulogen (**Fig. 11**) have the same distribution. Scale 20 μm. Ref.: [9]

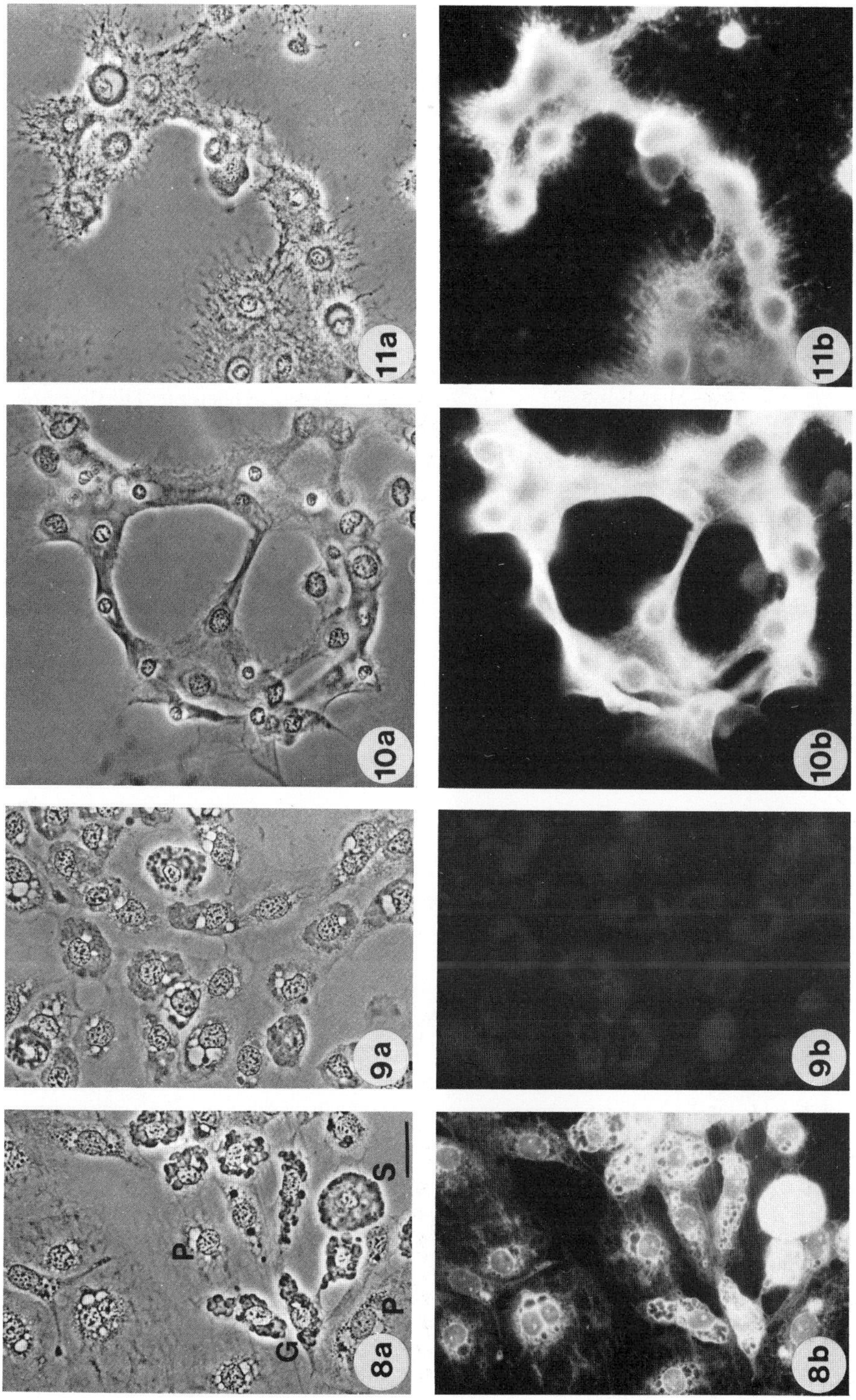
11a
11b
10a
10b
9a
9b
8a
8b
P
P
G
S

linking of the coagulogens might occur in various ways. Three representative models are put forward and discussed in more detail:

Model 1. Hemocyte and plasma coagulogen simultaneously build up a common network in which the two coagulogens – cross-linked with each other – might alternate in a more or less regular fashion.
Model 2. First a hemocyte gel is formed to which the plasma coagulogen then associates, filling the interspaces of the network. Subsequently, the network is stabilized by the formation of cross-links between the two coagulogens.
Model 3. A hemocyte gel is first established as in Model 2. The network of the gel is, however, only used as a matrix along which the plasma coagulogen molecules become arranged. Stabilization then takes place by cross-links formed only between the molecules of the plasma coagulogen.

Of the three models presented, No. 2 seems to be the most plausible. The first model is not in agreement with the observation that true clotting is still possible when the plasma coagulogen is added after a hemocyte gel has already formed (see Sects. 5.1 and 5.3). Moreover, one could expect that a clot – made up with the same amount of hemocytes as a gel – would occupy a larger volume because of the additional incorporation of the plasma coagulogen (Table 1). – Model 3, on the other hand, cannot satisfactorily explain why the hemocyte coagulogen develops resistence against DTT treatment during clotting.

5 Conditions of Clot Formation

5.1 Ions

Clotting presupposes the release of material from the hemocytes. Most of the substances which have proved to be potent inhibitors of hemolymph clotting [29,35] inhibit disintergration of the hemocytes and their granules. The most effective among them are chelating substances which bind Ca^{2+} and other divalent ions which means that Ca^{2+} is necessary for the rupture of the cells and their granules.

Ca^{2+} is, however, also required for the clotting processes starting after the release of the hemocyte material [8]. This was demonstrated by experiments in which rupture of the hemocytes was allowed to take place in the presence of Ca^{2+}. Immediately afterward, plasma containing EDTA (± specific ions) was added. Clotting is fully inhibited by relatively low concentrations of EDTA. The inhibition can be overcome by the addition of an excess of Ca^{2+}, Mn^{2+}, or Zn^{2+}; other divalent ions like Mg^{2+}, Be^{2+}, or Ba^{2+} were unable to replace Ca^{2+}.

5.2 Enzymes

In vertebrates, blood coagulation is brought about by the action of a series of activation factors working in succession. Most of them are serine proteases and dependent on the presence of Ca^{2+} for their activity or stability (for ref. see [38,41]).

The need for Ca^{2+} in hemolymph clotting of *Leucophaea* might indicate the activity of similar enzymes. Several natural trypsin inhibitors, as well as artificial inhibitors of serin proteases (like diisopropyl fluorophosphate), were tested, but did not show any effect on clotting. Clotting was, however, completely inhibited by cysteine blockers (like iodoacetamide) [8].

The iodoacetamide-sensitive factor is obviously derived from the hemocytes since prior treatment of plasma with iodoacetamide did not inhibit clotting. It is not clear whether the factor is identical with, or different from, the hemocyte coagulogen. Unfortunately, the rapid aging of the gel does not allow separation, separate treatment, and testing of gel and other soluble hemocyte factors without loss of clotting ability.

The inhibition of clotting by iodoacetamide could either mean that an SH enzyme is engaged in clotting, or that SH groups on the coagulogens are directly involved in cross-linking. The latter possibility can be ruled out. There are several observations which are not in agreement with the assumption of a cross-linking via disulfide bridges: (1) The clot does not dissolve in DTT; (2) iodoacetamide-treated plasma is still able to clot; (3) clotting is possible in the presence of DTT [8].

The activity of the supposed SH enzyme during clotting could consist of a limited proteolysis of the coagulogens as a prerequisite for their polymerization and cross-linking. Limited proteolysis is known from vertebrate fibrinogen and *Limulus* coagulogen (for ref. see [19]). In both cases, however, serin proteases are engaged which do not seem to play a role in *Leucophaea* clotting. Clotting of *Limulus* coagulogen can be artificially induced by treatment with trypsin [66]; in *Leucophaea*, treatment of the plasma coagulogen with trypsin or chymotrypsin is followed by the complete loss of its clotting activity (Bohn H, unpublished results). For the present, there are no indications of protolytic activity during hemolymph clotting.

The other possibility is that the SH enzyme is working as a cross-linking enzyme. The features of the SH enzyme are in good agreement with the characteristics given for the cross-linking enzymes of vertebrates, Crustacea [21,22] and Xiphosura [45]. They all are transglutaminases which need Ca^{2+} for their activation and which can be blocked by SH modifiers. The high similarity of the SH enzyme with the cross-linking enzymes would suggest that the enzyme is engaged in cross-linking rather than in proteolysis.

5.3 Hemocyte Coagulogen

It has already been mentioned (see Sect. 3.1.2) that the ability of the hemocyte coagulogen to make a clot together with the plasma coagulogen depends on its age, i.e., the time interval between its release and its contact with the plasma coagulogen. Within about 30 min there is an almost total loss of the clotting ability. To find out whether the loss of clottability is due to the aging of a soluble or of an insoluble component of the hemocytes, both components were separated by centrifugation and recombined after different aging. Normal clotting was only possible when the insoluble component, most likely the hemocyte gel, was fresh, irrespective of whether fresh or aged soluble hemocyte components were present. The aged gel, however, is still able to associate with plasma coagulogen (see Sect. 3.1.2).

The aging of the hemocyte gel and loss of clottability may be correlated with the establishment of new disulfide bridges observed during gel formation (see Sect. 3.2.2). Two explanations can be offered: (1) The development of new disulfide bridges could cause a conformational change in the gel after which a proper reaction with the plasma coagulogen for cross-linking is no longer possible; (2) by the formation of the disulfide bridges those important SH residues become blocked which have been shown to be important for cross-linking.

5.4 Cross-Linking

A preliminary attempt was made to find out how cross-linking is brought about. Clotting was studied in the presence of substances which are known to inhibit cross-linking of vertebrate and crustacean fibrin [22,49,54]. Glycine methyl ester, glycylglycine ethyl ester, Nα-p-tosyl-L-lysine methyl ester, and monodansylcadaverine did not affect clotting; glycineamide was slightly inhibitory; but hydroxylamine completely inhibited clotting [8,9a]. The effect of hydroxylamine might suggest that amino groups participate in cross-linking of *Leucophaea* coagulogens, similar to vertebrates and crustaceans, in which amino groups of lysine become linked with glutamyl residues. The importance of amino groups for the interaction of the coagulogens was further underlined by experiments in which all accessible amino groups of the plasma coagulogen had been modified by acetimidate. The modified plasma coagulogen is unable to clot [9a]. Whether in both cases cross-linking is affected or other aspects of the interaction of the coagulogens remains to be ascertained.

5.5 Lipid

The high content of lipid in the plasma coagulogen molecule raises the question of its participation in the clotting process. At least part of the lipid – the polar diglycerides – seems to be carried on the outer surface of the proteins [68] and might, therefore, interfere with clotting. The affinity of the two coagulogens for each other, for example, might be partly due to the interaction of the lipid moiety of the plasma coagulogen with a hydrophobic counterpart on the hemocyte coagulogen molecule.

Experiments with delipidated plasma coagulogen could possibly provide some information on the role of the lipid moiety during clotting, provided that the removal of the lipid does not cause a too severe denaturation of the plasma protein.

There is no doubt that the plasma coagulogen maintains its lipid moiety during clotting. The plasma coagulogen released after trypsin digestion of the clot still contains lipid (Bohn H., unpublished).

6 Conclusions

From the facts presented in the foregoing paragraphs the following picture of hemolymph clotting in *Leucophaea* may be drawn (Fig. 12). *Leucophaea* has two clotting

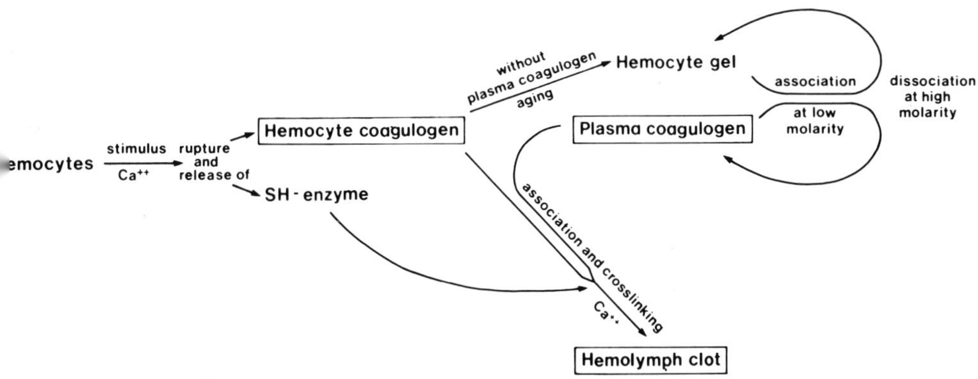

Fig. 12. Main reactions and factors taking part in hemolymph clotting of *Leucophaea*. Ref.: [8]

proteins, the hemocyte coagulogen, which is contained in the granules of the hemocytes, and the plasma coagulogen, which is contained in the hemolymph plasma. Induced by a chemical or mechanical stimulus clotting starts with the rupture of the hemocytes, provided that Ca^{2+} is present. The ruptured hemocytes release the hemocyte coagulogen and other substances into the surrounding hemolymph. Hemocyte and plasma coagulogen associate and become cross-linked to build the firm hemolymph clot. Cross-linking requires the presence of Ca^{2+} and the activity of a hemocyte-derived SH factor. In the absence of plasma coagulogen the hemocyte coagulogen also forms a coagulum, the hemocyte gel. The gel soon ages and loses the ability to cross-link with plasma coagulogen. But it can still associate with the plasma coagulogen. Association is especially effective at low ionic strength and can be reversed at high ionic strength.

The question arises as to whether this picture of clotting can be generalized for the insects. The great similarity between the plasma coagulogens of insects, as distantly related as are cockroaches and moths, would imply that clotting is very similar throughout the insects. It has indeed shown that the plasma coagulogen of *Periplaneta* (Blattoidea, Blattidae) can make full clotting together with the hemocytes of *Leucophaea* (Blaberoidea, Blaberidae).

On the other hand, clotting can be very different even between species of the same insect order. In the Coleoptera, for example, all four coagulation patterns may be found [34]. We have compared clotting in different cockroach species and also found considerable differences (Dantanello C. and Bohn H., unpublished). In the Blaberidae (*Leucophaea, Nauphoeta, Gromphadorhina, Blaberus*) the clotted material after centrifugation (15 min at 30 000 x g) makes up between 35 and 60% of the hemolymph volume, but in the Blattidae (*Periplaneta, Blatta, Eurycotis*) and Blattellidae (*Blattella*) it may be less than 10%, comparable to what is found in *Locusta* [11]. In correlation with the reduced clot volume much less plasma coagulogen is used during clotting, and a considerable amount of it remains in the serum. Undoubtedly, the hemocytes are responsible for these differences. *Periplaneta* hemocytes combined with *Leucophaea* plasma produce a meagre clot characteristic of *Periplaneta*; in the reverse combination a bulky clot is formed as in *Leucophaea*.

In spite of the differences between the clotting patterns, all show similarity in some of the features which would imply that the underlying basic pattern of clotting is the same in most insects. The constant element in the basic pattern, in my eyes, is the existence of two interacting coagulogens. What may be variable is the relative amounts and importance of the two coagulogens, the conditions under which they interact, and the type of their interaction.

A completely different concept of blood clotting was proposed by Gilliam and Shimanuki [26]. According to these authors the coagulation system of the honey bee is very similar to the human system: seven of the human clotting factors should be present in the hemolymph. Bee hemolymph was tested in a system developed for the study of human blood clotting factors. It may well be that hemolymph had some effect on the clotting of human blood, for example, through the activity of some kind of unspecific protease; but the observations do not allow any statement to be made about the clotting system of the bee.

7 Biological Importance of the Clotting Processes

Clotting processes are of great importance in homoeostasis (wound sealing) and in immune response. The homoeostatic mechanisms in insects work extremely rapidly and effectively. The rapid sealing of the wound can be attributed mainly to the features of the hemocyte coagulogen: its storage in the fragile hemocytes, its explosion-like ejection, and almost immediate gelation after extrusion. Whether the stabilizing effect of the incorporation of the plasma coagulogen is of great importance for wound sealing in *Leucophaea* may be doubted in view of the firmness the gel already has. But in other species, which possibly have a softer gel, the plasma coagulogen may be indispensible for homoeostasis.

The interaction of the two coagulogens during clotting can be regarded as an adaption to the open hemolymph system of insects and arthropods. If clotting of the plasma coagulogen would be initiated by a soluble factor released from the hemocytes, the danger would be that the activation factor rapidly spreads throughout the hemocoel and causes clotting at places where it is not wanted. The dependence of plasma coagulogen clotting on the presence of the more or less insoluble hemocyte coagulogen guarantees that clot formation is restricted to the wound site where the hemocytes disintegrate.

However important clotting mechanisms may be for the closure of wounds, the role they play in immune response is presumably much more important for the survival of the animals. In various kinds of immune reactions of insects, such as phagocytosis, nodule formation, and encapsulation (for ref. see [58]), the hemocytes degranulate and release clotting substances which entrap or coat foreign particles to make them suitable for elimination. One could imagine that the tightening of the coagulum structure by the plasma coagulogen is necessary to render the meshwork impassable for microorganisms.

References

1. Akai H, Sato S (1973) Ultrastructure of the larval hemocytes of the silkworm, *Bombyx mori* (L) (Lepidoptera: Bombycidae). Int J Insect Morphol Embryol 2: 207-231
2. Arnold JW (1974) The hemocytes of insects. In: Rockstein M (ed) The physiology of insecta, 2nd edn, vol 5. Academic Press, New York, pp 201-254
3. Barwig B (1985) Isolation and characterization of plasma coagulogen (PC) of the cockroach *Leucophaea maderae* (Blattaria). J Comp Physiol 155: 135-143
4. Barwig B, Bohn H (1980) Evidence for the presence of two clotting proteins in insects. Naturwissenschaften 67: 47
5. Beard RL (1950) Experimental observations on coagulation of insect hemolymph. Physiol Zool 23: 47-57
6. Bohn H (1977) Differential adhesion of the haemocytes of *Leucophaea maderae* (Blattaria) to a glass surface. J Insect Physiol 23: 185-194
7. Bohn H (1977) Enzymatic and immunological characterization of the conditioning factor for epidermal outgrowth in the cockroach *Leucophaea maderae*. J Insect Physiol 23: 1063-1073
8. Bohn H, Barwig B (1984) Hemolymph clotting in the cockroach *Leucophaea maderae* (Blattaria). Influence of ions and inhibitors; isolation of the plasma coagulogen. J Comp Physiol 154: 457-467
9. Bohn H, Barwig B, Bohn B (1981) Immunochemical analysis of hemolymph clotting in the insect *Leucophaea maderae* (Blattaria). J Comp Physiol 143: 169-184

9a. Bohn H, Saks TM (1986) Inhibition of crosslinking in clotting of cockroach hemolymph (*Leucophaea maderae*) (Blaberidae, Blattaria) J Comp Physiol (in press)

10. Bowen TJ, Kilby BA (1953) Electrophoresis of locust haemolymph. Arch Int Physiol Biochim 61: 413-416
11. Brehélin M (1971) La coagulation de l'hémolymphe et ses modalités chez les larves de l'Orthoptère *Locusta migratoria* (phase grégaire). C R Acad Sci Paris Sér D 273: 1598-1601
12. Brehélin M (1972) Étude de mécanisme de la coagulation de l'hémolymphe d'un Acridien: *Locusta migratoria migratorioides* (R and F) Acrida 1: 167-175
13. Brehélin M (1979) Hemolymph coagulation in *Locusta migratoria*: Evidence for a functional equivalent of fibrinogen. Comp Biochem Physiol 62B: 329-334
14. Brehélin M (1979) Mise en évidence de l'induction de la coagulation plasmatique par les hémocytes chez *Locusta migratoria*. Experientia (Basel) 35: 270
15. Chino H, Kitizawa K (1981) Diacylglycerol-carrying lipoprotein of hemolymph of the locust and some insects. J Lipid Res 22: 1042-1052
16. Chino H, Murakami S, Harashima K (1969) Diglyceride-carrying lipoproteins in insect hemolymph. Isolation, purification and properties. Biochim Biophys Acta 176: 1-26
17. Chino H, Downer RGH, Wyatt GR, Gilbert LI (1981) Lipophorins, a major class of lipoproteins of insect haemolymph. Insect Biochem 11: 491
18. Chino H, Katase H, Downer RGH, Takahashi K (1981) Diacylglycerol-carrying lipoprotein of hemolymph of the American cockroach: Purification, characterization, and function. J Lipid Res 22: 7-15
19. Cohen E (ed) (1979) Biomedical applications of the horseshoe crab (Limulidae). Liss, New York (Progr Clin Biol Res, vol 29)
20. Crossley ACS (1975) The cytophysiology of insect blood. Adv Insect Physiol 11: 117-222
21. Doolittle RF, Lorand L (1962) Inhibitors of lobster blood clotting. Biol Bull (Woods Hole) 123: 481-482
22. Fuller GM, Doolittle RF (1971) Studies of invertebrate fibrinogen II. Transformation of Lobster fibrinogen into fibrin. Biochemistry 10: 1311-1315
23. Gellissen G (1983) Lipophorin as the plasma coagulogen in *Locusta migratoria*. Naturwissenschaften 70: 45-46
24. Gellissen G, Emmerich H (1980) Purification and properties of a diglyceride-binding lipoprotein (LP I) of the hemolymph of adult male *Locusta migratoria*. J Comp Physiol 136B: 1-9
25. Gellissen G, Wyatt GR (1981) Production of lipophorin in the fat body of adult *Locusta migratoria*: comparison with vitellogenin. Can J Biochem 59: 648-654
26. Gilliam M, Shimanuki H (1970) Coagulation of hemolymph of the larval honey bee (*Apis mellifera* L). Experientia (Basel) 26: 908-909

27. Goffinet G, Grégoire C (1975) Coagulocyte alterations in clotting hemolymph of *Carausius morosus* L. Arch Int Physiol Biochim 83: 707-722
28. Goffinet G, Grégoire C (1979) Gross and ultrastructural changes in coagulocytes of mole-cricket, *Gryllotalpa* sp. (Orthoptera: Gryllotapidae) during in vitro hemolymph coagulation. Int J Insect Morphol Embryol 8: 59-66
29. Grégoire C (1953) Blood coagulation in Arthropods III. Reactions of insect hemolymph to coagulation inhibitors of vertebrate blood. Biol Bull (Woods Hole) 104: 372-393
30. Grégoire C (1955) Blood coagulation in Arthropods V. Studies on hemolymph coagulation in 420 species of insects. Arch Biol 66: 103-148
31. Grégoire C (1970) Hemolymph coagulation in arthropods. Symp Zool Soc Lond 27: 45-74
32. Grégoire C (1971) Hemolymph coagulation in Arthropods. In: Florkin M, Scheer BT (eds) Chemical zoology, vol 6, pp 145-189
33. Grégoire C (1974) Hemolymph coagulation. In: Rockstein M (ed) The physiology of insecta, 2nd edn, vol 5, pp 309-360
34. Grégoire C, Florkin M (1950) Blood coagulation in Arthropods I. The coagulation of insect blood, as studied with the phase contrast microscope. Physiol Comp Oecol 2: 126-139
35. Grégoire C, Goffinet G (1979) Controversies about the coagulocyte. In: Gupta AP (ed) Insect hemocytes. Cambridge University Press, Cambridge, pp 189-229
36. Gupta AP (1979) Arthropod hemocytes and phylogeny. In: Gupta AP (ed) Arthropod phylogeny. New York, pp 669-735
37. Hoffmann JA, Stoeckel M-E (1968) Sur les modifications ultrastructurales des coagulocytes au cours de la coagulation de l'hémolymphe chez un Insecte Orthoptéroides: *Locusta migratoria* L. C R Soc Biol 162: 2257-2259
38. Jackson CM, Nemerson Y (1980) Blood coagulation. Annu Rev Biochem 49: 765-811
39. Katase H, Chino H (1982) Transport of hydrocarbons by the lipophorin of insect hemolymph. Biochim Biophys Acta 710: 341-348
40. Katase H, Chino H (1984) Transport of hydrocarbons by haemolymph lipophorin in *Locusta migratoria*. Insect Biochem 14: 1-6
41. Kazal LA (1982) Coagulation chemistry. Clin Biochem Contemp Theories Techn 2: 73-143
42. Kunkel JG, Nordin JH (1985) Yolk proteins. In: Kerkut GA, Gilbert LI (eds) Comprehensive Insect Physiol Biochem Pharmacology. Pergamon, Oxford, vol 1, pp 83-111
43. Kunkel JG, Pan ML (1976) Selectivity of yolk protein uptake: Comparison of vitellogenins of two insects. J Insect Physiol 22: 809-818
44. Levin J, Bang FB (1964) A description of cellular coagulation in the *Limulus*. Bull J Hopkins Hosp 115: 337-345
45. Liu T-Y, Seid RC, Tai JY, Liang S-M, Sakmar TP, Robbins JB (1979) Studies on *Limulus* lysate coagulating system. Progr Clin Biol Res 29: 147-158
46. Loeb L (1903) On the coagulation of the blood of some arthropods and on the influence of pressure and traction on the protoplasm of the blood cells of arthropods. Biol Bull (Woods Hole) 4: 301-318
47. Loeb L (1904) Über die Koagulation des Blutes einiger Arthropoden. Beitr Chem Physiol Pathol 5: 191-207
48. Loeb L (1910) Über die Blutgerinnung bei Wirbellosen. Biochem Z 24: 478-495
49. Lorand L, Doolittle RF, Konishi K, Riggs SK (1963) A new class of blood coagulation inhibitors. Arch Biochem Biophys 102: 171-179
50. Marschall KJ (1966) Bau und Funktionen der Blutzellen des Mehlkäfers *Tenebrio molitor* L. Z Morphol Ökol Tiere 58: 182-246
51. Moran DT (1971) The fine structure of cockroach blood cells. Tissue **Cell** 3: 413-422
52. Muttkowski RA (1924) Studies on the blood of insects III. The coagulation and clotting of insect blood. Bull Brooklyn Entomol Soc 19: 128-144
53. Neuwirth M, Lai-Fook J (1977) Cytochemistry of the very early stage of in situ coagulation in *Calpodes ethlius* (Lepidoptera). Can J Zool 55: 1767-1772
54. Nilsson JLG, Stenberg P, Ljunggren C, Hoffman K-J, Lunden R, Eriksson O, Lorand L (1972) Fibrin-stabilizing factor inhibitors. Ann N Y Acad Sci 202: 286-296

55. Pattnaik NM, Mundall EC, Trambusti BG, Law JH, Kézdy FJ (1979) Isolation and characterization of a larval lipoprotein from the hemolymph of *Manduca sexta*. Comp Biochem Physiol 63B: 469-476
56. Peled Y, Tietz A (1975) Isolation and properties of a lipoprotein from the haemolymph of the locust, *Locusta migratoria*. Insect Biochem 5: 61-72
57. Ratcliffe NA, Gagen SJ (1977) Studies in the in vivo cellular reactions of insects: An ultrastructural analysis of nodule formation in *Galleria mellonella*. Tissue Cell 9: 73-85
58. Ratcliffe NA, Rowley AF (1979) Role of hemocytes in defense against biological agents. In: Gupta AP (ed) Insect hemocytes. Cambridge University Press, Cambridge, pp 331-414
59. Ravindranath MH (1980) Haemocytes in haemolymph coagulation of arthropods. Biol Rev 55: 139-170
60. Rowley AF (1977) The role of the haemocytes of *Clitumnus extradentatus* in haemolymph coagulation. Cell Tissue Res 182: 513-524
61. Rowley AF, Ratcliffe NA (1976) The granular cells of *Galleria mellonella* during clotting and phagocytic reactions in vitro. Tissue Cell 8: 437-446
62. Rowley AF, Ratcliffe NA (1981) Insects. In: Ratcliffe NA, Rowley AF (eds) Invertebrate blood cells, vol 2. Academic Press, New York, pp 421-490
63. Scharrer B (1972) Cytophysiological features of hemocytes in cockroaches. Z Zellforsch Mikrosk Anat 129: 301-319
64. Shapiro JP, Keim PS, Law JH (1984) Structural studies on lipophorin, an insect lipoprotein. J Biol Chem 259: 3680-3685
65. Siakotos AN (1960) The conjugated plasma proteins of the American cockroach II. Changes during the molting and clotting process. J Gen Physiol 43: 1015-1030
66. Solum NO (1973) The coagulogen of *Limulus polyphemus* hemocytes. A comparison of the clotted and non-clotted forms of the molecule. Thromb Res 2: 55-70
67. Stang-Voss C (1970) Zur Ultrastruktur der Blutzellen wirbelloser Tiere I. Über die Haemocyten der Larve des Mehlkäfers *Tenebrio molitor*. Z Zellforsch Mikrosk Anat 103: 589-605
68. van der Horst DJ, van Doorn JM, de Keijzer AN, Beenakkers AMT (1981) Interconversions of diacylglycerol-transporting lipoproteins in the haemolymph of *Locusta migratoria*. Insect Biochem 11: 717-723
69. Wheeler CH, Mundy JE, Goldsworthy GJ (1984) Locust haemolymph lipoproteins visualised in the electron microscope. J Comp Physiol 154B: 281-286
70. Yeager JF, Knight HH (1933) Microscopic observations on blood coagulation in several different species of insects. Ann Entomol Soc Am 26: 591-602
71. Yeager JF, Shull WE, Farrar MD (1932) On the coagulation of blood from the cockroach, *Periplaneta orientalis* (Linn), with special reference to blood smears. Iowa State Coll J Sci 6: 325-344

Chapter 15 The Prophenoloxidase Activating System: The Biochemistry of Its Activation and Role in Arthropod Cellular Immunity, with Special Reference to Crustaceans

K. SÖDERHÄLL[1] and V. J. SMITH[2]

CONTENTS

1 Introduction

Although modern immunology owes its origin largely to the pioneering work of Metchnikoff [40] and his observations of phagocytosis in the water flea, *Daphnia*, the study of "immune" processes in invertebrates has always lagged behind that of mammals. Indeed, it is only recently that plausible theories for the discrimination of self from non-self in invertebrates have been put forward and that fundamental techniques such as the separation of the various blood cell populations have been devised. As it is, there is still debate between invertebrate immunologists as to the true biochemical/molecular basis for non-self recognition in different invertebrate phyla, and foremost amongst the mechanisms proposed to account for recognition phenomena in the arthropods are the lectins, cell surface charge and, more recently, the prophenoloxidase activating system. To help clarify the picture for arthropods, the present review aims to re-evaluate some of the earlier work conducted on arthropod cellular immunity which gave rise to these concepts, and to reinterpret the data provided by the respective authors in the light of our present understanding of the prophenoloxidase activating system in crustaceans.

1 Institutionen för Fysiologisk Botanik, Uppsala Universitet, Box 540, 751 21 Uppsala, Sweden
2 University Marine Biological Station Millport, Isle of Cumbrae, Scotland KA28 OEG, U.K.

Immunity in Invertebrates
Ed. by M. Brehélin

1.1 Concepts of Immunorecognition

Much of the early research into the invertebrate immune system focussed on the search for antibody-type recognition factors or lymphocyte-like cells within the body fluids. It is now generally accepted that invertebrates do not express immunoglobulins and do not contain T lymphocytes as part of their haemograms, but the notion that particular recognition factors reside in invertebrate blood has persisted and popular candidates for these factors have been the agglutinins [1,11,77,82]. The ability of invertebrates to agglutinate foreign cells, such as mammalian erythrocytes, has been known since 1903, when Noguchi [42] reported the agglutination of red blood cells by the haemolymph of the horse-shoe crab, *Limulus polyphemus*. Haemagglutinins have now been detected in a wide range of organisms, including plants [31,32], vertebrates [33] and many lower animals (see review by Yeaton [84]), and several have been purified and characterized (see Yeaton [85]). For the majority of species, these agglutinins, or more correct lectins, comprise a heterogenous group of glycoproteins or proteins that are heat-labile, and have a polyvalent configuration which binds specifically to cell surface polysaccharides [84,85]. Agglutinating activity is inhibited by competitive binding with particular sugars [85], and in some cases has been shown to be calcium dependent [34].

The belief that haemagglutinating molecules might serve in non-self recognition in invertebrates stems from the early in vitro investigations of cellular defence reactions, such as that of McKay, Jenkin and Tyson on the Australian crayfish, *Parachaeraps bicarinatus* (now renamed *Cherax destructor*) and Paterson and Stewart [43] and Stewart and Zwicker [74] on the American lobster, *Homarus americanus*. In a classic series of papers, McKay et al. [38], McKay and Jenkin [36,37] and Tyson and Jenkin [78, 79] reported that the phagocytosis and elimination of erythrocytes or bacteria by the haemocytes of *C. destructor* in vitro or in vivo was dependent upon recognition factors present in the haemolymph (sera) or on the cell membranes of the host reticulo-endothelial network. The authors speculated that the haemagglutinins in sera of *C. destructor* might be involved in the recognition process as these molecules displayed quite distinct specificity in action and could be absorbed away with consequent loss of opsonic effect [36,38,79]. Resistance to infection against a crayfish pathogen, *Pseudomonas* CP, was similarly linked to serum-dependent phagocytosis, and short-lived enhancement of resistance was reported to be induced in *C. destructor* by injections of low doses of lipopolysaccharides (LPS) [37]. Likewise with the American lobster, Stewart and Zwicker [74], Cornick and Stewart [15] and Paterson and Stewart [43] considered serum recognition factors, possibly related to the agglutinins, to be important in cellular responsiveness, although, in other arthropods, Scott [55], Anderson [2], Smith and Ratcliffe [56], McCumber et al. [35] and Rowley and Ratcliffe [53] failed to detect opsonic activity in the sera or haemolymph. Clearly the role of haemagglutinins in arthropod immunity is a matter of controversy and a few workers, principally Yoshino [86] and Renwrantz and Stahmer [49] have undertaken detailed analyses of invertebrate lectins to resolve their role in immunorecognition and immunocompetence in these animals. As yet, it is still unclear how the lectins operate in vivo and the nature and extent of any regulatory mechanisms which modulate their reactivity in cellular resistance. Neither is it known why some arthropod species possess multiple lectins within the blood whilst others appear to have none, despite the universal need for arthropods (and indeed all invertebrates) to protect themselves against microbial or parasitic invasion. Certainly, immunization of arthropods with bacteria, yeasts or their cell wall products, confers temporary immunity on the host, as shown for example by Stephens [73], Evans et al. [17], McKay and Jenkin [37] and Hultmark et al. [25], but how this relates to lectin-mediated recognition remains an enigma. Komano et al. [28] have succeeded in inducing lectins specific for D-galactose in *Sarcophaga peregrina* larvae by injury to the body wall, but in many cases, immunization has been found to lead to elevated bactericidal or lysozyme activity in the blood, and the ways this is achieved in most species is unknown.

An alternate theory, put forward to account for differential haemocytic responsiveness to non-self entities is that of the surface charge of the challenge material which may or may not favour haemocyte attachment in vivo [29]. Unfortunately, convincing experimental evidence that differences in surface charge effect haemocyte behavior in vitro has not been forthcoming, and in mammals, surface charge and van der Waals forces are no longer considered to play a significant role in cell adhesion during tissue differentiation and development [16].

In contrast, evidence is rapidly accumulating that the prophenoloxidase activating system is important in arthropod cellular defence and the data favouring this supposition has been reviewed by us previously [61,66,67].

2 The Prophenoloxidase Activating System: Biochemical Aspects

Briefly, the prophenoloxidase activating system is a complex cascade of enzymes and other factors that is ultimately responsible for the initiation of melanin synthesis by the host. That melanization is involved in cellular immunity in arthropods has been mooted for some time, particularly as melanin deposition is frequently observed to accompany the host cellular responses to wounding and parasitic or microbial invasion [54,76]. However, the exact relationship between melanin synthesis and non-self recognition remained a matter of speculation until Pye [46] and Unestam and Söderhäll [81] demonstrated that yeasts [46,81] or their constituent β1,3-glucans [81] initiated melanization reactions in arthropod haemolymph by activating prophenoloxidase, the key enzyme in the melanization reactions. It is now well established that not only β1, 3-glucans, but also bacterial cell walls or the lipopolysaccharides borne on these structures specifically trigger activation of the proPO system, and the process involves an entire orchestra of intermediary proteins [6,60,62,64,68].

2.1 Constituent Factors of the Prophenoloxidase Activating System

Phenoloxidase itself represents the terminal component of the proPO system [61] (Fig. 1). In most arthropods the enzyme occurs more usually in the form of the inactive proenzyme, prophenoloxidase, and conversion of prophenoloxidase to phenoloxidase is brought about by a limited proteolysis [6,60] (Fig. 1). The sequence of

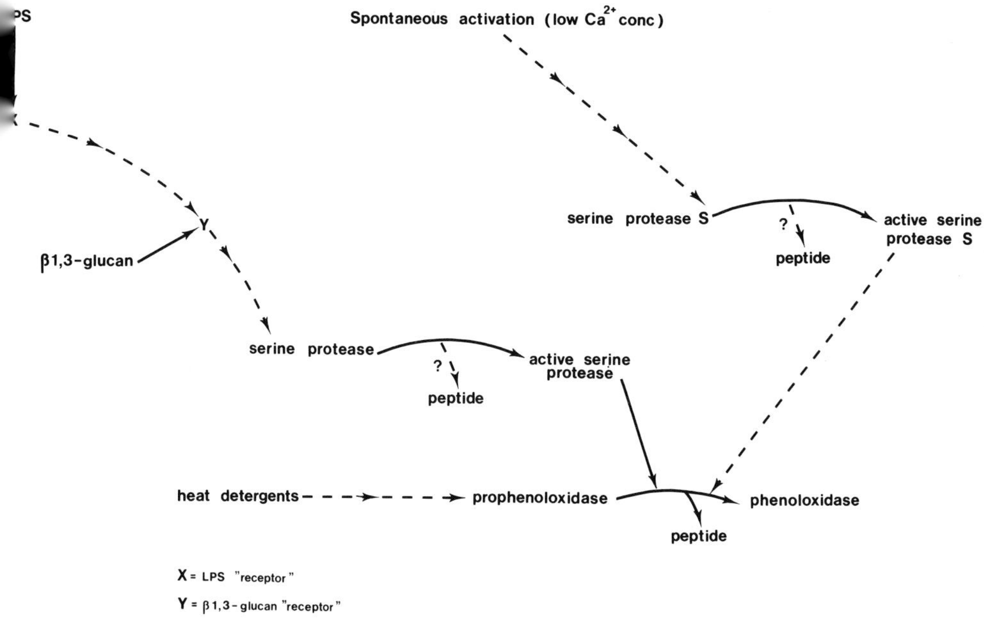

Fig. 1. The crustacean prophenoloxidase activating system. *Dashed lines* denote inferred pathways

events which mediate prophenoloxidase activation has been most extensively worked out for the freshwater crayfish, *Astacus astacus* (see reviews by Söderhäll [61], Söderhäll and Smith [67]) and the silk moth, *Bombyx mori* [3,4,5,7], but preliminary research on crabs and insects [30,57] has indicated that the processes are shared by several groups of arthropods and are probably universal for the phylum as a whole. Amongst the constituent factors of the proPO cascade are a number of serine proteases (Fig. 1) [62]; and of these, one, the activating enzyme, is most likely a zymogen requiring preactivation by proteolysis for activity [6,62]. The enzyme is dependent on calcium and is highly specific in action, hydrolyzing only peptides with the structure R_1-Gly-Arg-R_2 [62]. The enzyme is very sensitive to stimulation by non-self molecules and is activated, through an as yet undefined route, by glucan or LPS concentrations as low as 10^{-10} g ml^{-1} [64]. Curiously in crayfish, LPS levels in excess of 10^{-4} g ml^{-1} actually block prophenoloxidase activation although the reason for this is unclear.

A second serine protease, designated protease S, similarly effects the conversion of prophenoloxidase to phenoloxidase, but while it demonstrates the same hydrolytic specificity as the activating enzyme, it differs from it in not requiring LPS or β1,3-glucans for activation [64]. Instead, protease S is triggered by low calcium concentrations (< 5 mM), and in common with other cascade components, most likely generates a small peptide after proteolytic cleavage. This, however, remains to be demonstrated.

Other serine proteases also reside within the proPO pathway and some of them may constitute, together with phenoloxidase, the five sticky or attaching proteins which are formed upon induction of the cascade by glucans of LPS [69] (Fig. 1). Little is known about the biochemistry or biological function of these attaching proteins, however, and an evaluation of their characteristics is urgently needed. Information is also wanting on the "receptor" by which the non-self carbohydrates switch on the proPO cascade. So far, all we know is that in crayfish separate receptors exist for the glucans and LPS within the system as haemocyte lysate supernatants (HLS) containing a proPO preparation inhibited by high levels of LPS can still be activated by β1,3-glucans [64]. However, it is unknown whether the cascade is stimulated by different non-self molecules (e.g. lipoteichoic acids or peptidoglycans from Gram positive bacteria) through receptors separate again from those for glucans or LPS and in the case of protozoan or helminth parasites, the nature of the eliciting signal is still a mystery. As far as nonphysiological inducers of proPO (e.g. heat: 58°C 10 min [6]), detergents [27] or organic solvents [45] are concerned, it is unlikely that particular receptors occur within the cascade, since only the last component proPO is activated by these agents and activation through these agents probably involves conformational change of the proPO. However, the ways in which lipids [24] bring about activation of prophenoloxidase is unclear.

Perhaps better understood, are the regulatory factors which exist in the system. For instance, naturally occurring proteolytic inhibitors have been identified in the haemocytes and cuticle of *A. astacus* [21,22] and while the precise role of these inhibitors in proPO activation awaits confirmation, it is likely that they serve to modulate the activating reactions in some way. In addition, Hergenhahn and Söderhäll [23] have found α_2 macroglobulin-like activity in the blood of crayfish which might serve to sequester proteases from the circulation in a manner analogous to α_2-macroglobulin of mammals [71]. Modulation of proPO activation might be further achieved through the time-dependent self-inactivation of the serine protease which activates prophenoloxidase [62]. The activity of this protease is lost if glucan-treated crayfish haemocyte lysate (HLS) preparations are incubated for ca. 60 min at 37°C or left to stand for 4–5 h at room temperature [62].

Lastly, amongst the constituent factors of the proPO pathway are the proteins involved in coagulation (Fig. 1). A connection between coagulation and proPO activation was demonstrated by Söderhäll in 1981 [60] when HLS preparations exposed to β1, 3-glucans were seen to undergo a gelation or clotting reaction. The response has been found to be calcium dependent and to be mediated by proteolysis [60]. The protease responsible for this exhibits identical specificity as the activating enzyme (see above) and could therefore be linked to the proPO system [60]. At present, the full repertoire of proteins which participate in the conversion of the coagulogen to the coagulin and thence to the final clot has not been ascertained, but the clotting reaction of crayfish shares several characteristics with the clotting reactions of the *Limulidae* and to certain factors in the mammalian system (see Söderhäll and Smith [67]) and thus could represent an evolutionary ancient pathway which has been conserved during phylogeny.

2.1.1 Activating Pathways

Interestingly, the proPO pathway of crustaceans mimicks not only the clotting cascade of *Limulus* but also the complement pathway of vertebrates [66,67]. In particular, it possesses two separate routes through which activation can proceed. One of these, the "alternate" pathway, operates upon induction with non-self molecules and involves the activating enzyme and at least one other serine protease [6,64] (Fig. 1). In contrast, the other pathway, the "classical" pathway, operates independently of foreign molecules and is provoked by low levels of calcium (< 5 mM) (see above) [64] (Fig. 1). Functionally, the alternate pathway of proPO activation is probably used when microbial invasion threatens the host homeostatic integrity, while the classical pathway functions to "prime" the blood cells and invoke coagulation when the host sustains some injury or blood loss through the cuticular surface.

2.1.2 Occurence and Distribution of Proteins of the Prophenoloxidase System

For the majority of arthropods, the proPO system resides in the haemocytes [30,57, 64,65,68]. In a few species of insects the proPO cascade is thought to reside in the plasma [4], but, as we have discussed in a separate publication [66], this could well be a consequence of the method used to obtain the haemolymph samples from the test organism, particularly as some insect haemocytes are known to be exceptionally sensitive to handling or ionic shock [47]. An unusual situation seems to occur in the *Limulidae*, where the terminal component of the cascade, phenoloxidase, is lacking from the amoebocytes, while other constituent proteins are present (Söderhäll K., Levin J., Armstrong P., in prep.). The extent to which this reflects the taxonomic position of the *Limulidae* is open to question, but members of the Echinodermata, Tunicata and Mollusca have all been found to exhibit phenoloxidase activity in the blood (Söderhäll K. and Bertheussen K., unpublished; Smith V.J. and Söderhäll K., in prep.). More careful analyses of phenoloxidase activity with separated cell populations has further revealed that for most species the phenoloxidase is present in the form of a proenzyme and occurs in the granular and semigranular type cells (Söderhäll K. and Bertheussen K., unpublished) in a similar way to that demonstrated by crustaceans ([65], Smith V.J. and Söderhäll K., in prep.). Unfortunately, little is known about the biochemical pathways which lead to phenoloxidase activation in these non-arthropod invertebrates, or of the sensitivity of the enzymes to activation by non-self molecules. Once again, research into these and other features of the proPO system in lower animals is required.

3 The Prophenoloxidase Activating System and Host Cellular Immunity

While the relationship between non-self recognition and phenoloxidase activity was first demonstrated by Pye [46], Unestam and Söderhäll [81] and Söderhäll and Unestam [68] in insects and crustaceans respectively, the significance of the activating cascade in cellular responsiveness was enigmatic until Smith and Söderhäll [57] showed that elevated rates of phagocytosis of bacteria by crayfish or crab haemocytes in vitro could be obtained by inclusion of β1, 3-glucans in the cell: bacteria mixtures (Fig. 2). That it is the attaching proteins of the proPO cascade which possess opsonic function has now been confirmed ([67], Söderhäll K., Smith V.J., and Johansson M., in prep.) although, interestingly, phenoloxidase itself does not appear to promote ingestion of bacteria as isolated crab hyaline cells ingest bacteria, pretreated with heat activated HLS (in which only phenoloxidase is generated) at the same rate as control, saline-incubated bacteria, whereas bacteria opsonized with glucan-activated HLS are avidly phagocytosed. Correlative in vivo investigations with crayfish have also revealed that stronger encapsulation reactions are shown towards fungal spores coated in crayfish HLS than towards fungal spores preincubated in plasma or buffer [69] and, in crabs, nodule formation has been induced in the haemocoel specifically by injection of dissolved β1, 3-glucans (but not other dissolved glycans) in the absence of any particulate material [59]. An opsonic role for the proPO system has similarly been reported for insects [48], but in neither crustaceans nor insects have the precise proPO proteins which mediate the response been characterized.

3.1 Opsonization and Immune Adherence

Notwithstanding, it is possible, in the light of our present understanding of proPO activation, to reassess the findings of McKay and his co-workers on *Cherax destructor* (see above). Certainly, the enhanced ingestion of "serum-opsonized" erythrocytes or bacteria by the haemocytes seen by these authors [36,78,79] can be attributed to the proPO system within the hosts haemocytes, as the sera used to pretreat the test particles were prepared from whole, clotted, crayfish haemolymphs [36] and clotting in crustaceans is known to involve the activation of the proPO cascade [60]. However, it is unlikely that the opsonic factors were derived totally from the sera as the self-inactivation of proPO components [62] or stickiness of the activated proteins [69] might have rendered them unavailable to the host phagocytes. Instead, in *C. destructor*, the sera might have caused release of the hosts own proPO-proteins (i.e. opsonins) in situ on the monolayers by stimulating exocytosis of the crayfish granular cells. Equivalent rapid exocytosis of crayfish granular cells has been found to be achieved by activated sticky proPO proteins in vitro by Smith and Söderhäll [58] and furthermore, the bacteria alone or LPS were able only to induce discharge of granules from the semigranular cells, leaving the granular cells unaffected (Johansson M. and Söderhäll K., in prep.) (Fig. 2). In crustaceans, the granular cells are the main repositories of the proPO cascade [65], so direct stimulation of these cells by the active proteins of the proPO cascade results in greater amounts of opsonins to be produced in the cultures. Presumably, with *C. destructor,* the serum treatment of the bacteria [78,79] stimulated greater and more rapid release of proPO proteins from the granular cells and more rapid release of proPO proteins from the granular cells on the coverslips than the bacteria could achieve by themselves and the attaching proteins thus generated in the cultures would then be available to the phagocytes. The production of "self" proPO proteins

Fig. 2a-d. Monolayers of separated crayfish haemocytes. **a,b** Semigranular cells: **a** unaffected cells; **b** degranulated (= exocytosed cells, after incubation with LPS 100 μg ml^{-1}. **c,d** Granular cells: **c** unaffected cells; **d** degranulated (= exocytosed) cells after incubation with an active haemocyte lysate (active proPO proteins). Nomarski differential interference contrast microscopy, *bars* 10 μM. (From Mats W. Johansson)

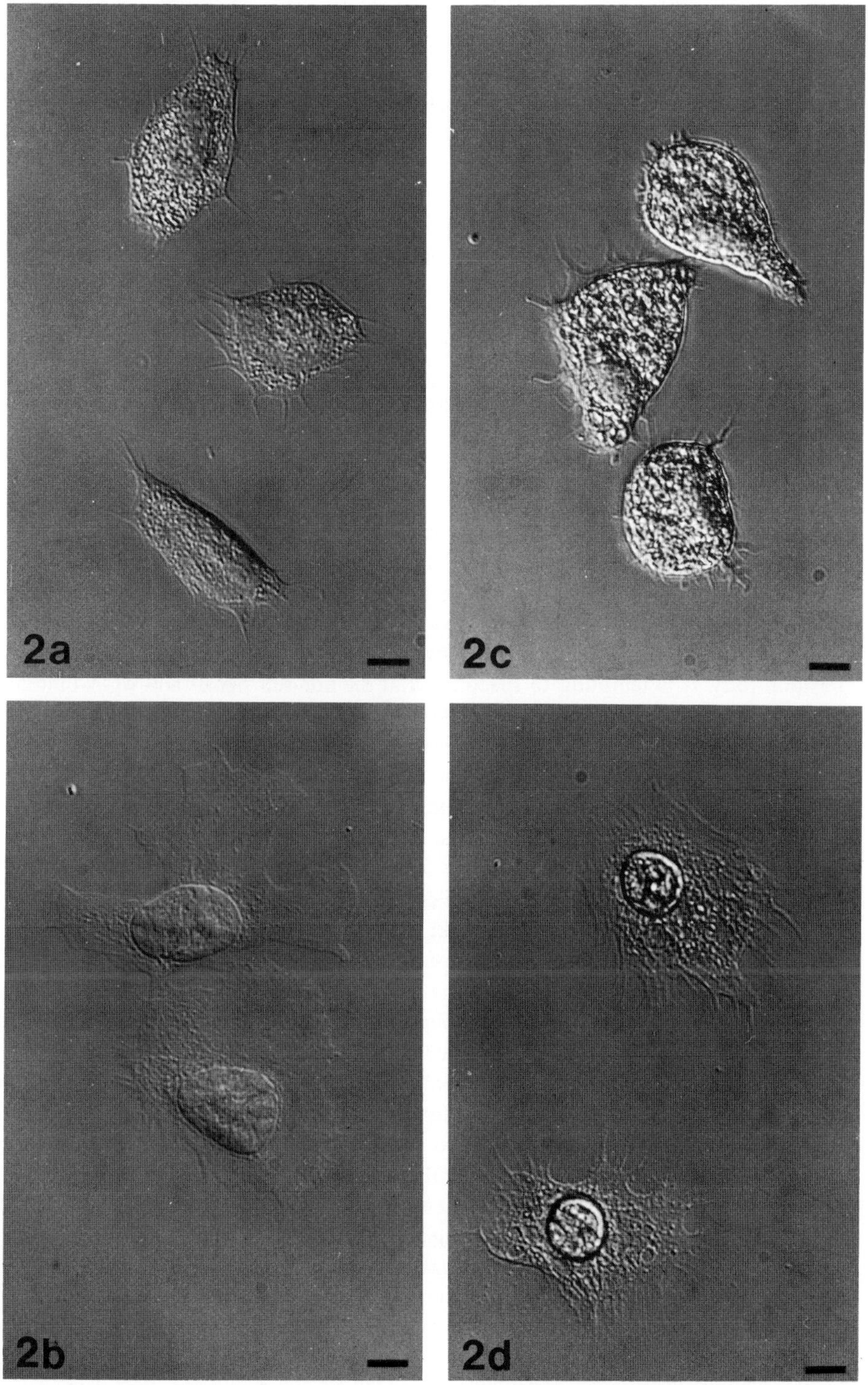
2a
2c
2b
2d

from the granulocytes would also account for the apparent presence of recognition factors on, or in association with, the cell membrane of the effector haemocytes [78], and the loss of opsonic activity upon trypsin treatment [36] was probably due to the total destruction of the putative opsonins before interaction with the monolayer cells. Moreover, in view of the lability of arthropod haemocytes to disruption during experimental manipulation, great caution should be exercised when setting up monolayer cultures, and whenever possible separated haemocyte populations should be used to avoid inadvertent activation of proPO proteins in vitro. Cell separation techniques could also be applied to the analysis of capsule/nodule formation in vitro, although Rizki and Rizki [50,52] have employed the novel approach of using melanotic tumour formation in temperature-sensitive mutant strains of *Drosophila melanogaster* to study the processes in insects (see Chap. 13). In this work, Rizki and Rizki [50] identified the lamellocytes as the principal cells in tumour (or capsule) formation. They appear to differentiate from plasmatocytes but form two distinct sub-populations according to their affinity for wheat germ agglutinin (WGA) [52]. One of these sub-populations, the strong WGA-binding, or spk+ cells, predominates over the other, weak WGA-binding (spk-) cells during tumour (or capsule) development and participates actively in the cellular response [52]. The authors propose that the spk+ cells are stimulated to differentiate by a diffusible recognition signal released from the aberrent (non-self) tissue or material, and the WGA-binding substance expressed by the spk+ cells as a consequence of differentiation then serve to coat the abnormal (or foreign) surface, thereby creating a recognition site for subsequent spk+ and later spk- lamellocyte aggregation [52]. It is tempting to speculate that the non-self signals and WGA-binding materials described by these authors might be proteins with the proPO system, as the hypothetical events suggested by these authors are essentially the same as those observed with proPO components in crustaceans. However, in *D. melanogaster*, phenoloxidase activity is thought to reside within the crystal cells [50]. Whether other proPO enzymes, apart from phenoloxidase, exist in the plasmatocytes/lamellocytes in fruit flies is unknown, but even in the absence of such components from the lamellocytes, it is possible that these factors might be contributed to the spk+ cells through some form of cell cooperation. Rizki and Rizki [51] have further shown that tumour (or capsule) formation in *D. melanogaster* is suppressed by a factor produced by the eggs of the parasitic cynipid wasp, *Leptopilina heterotoma*. Inhibition seems to occur by impairment of the acquisition of adhesiveness of the lamellocytes [51] and it would be interesting to establish whether this factor from *L. heterotoma* females has any influence on haemolymph proPO activation in *D. melanogaster*. A factor from hymenopteran parasites was recently found to inhibit proPO activation in *Heliothis virescens* and *Trichoplausia ni* [75]. Returning to *C. destructor*, the enhanced clearance of "opsonized" test particles from the circulation observed by Tyson and Jenkin [78] may, likewise, be due to the proPO system. In this case, the non-self signals (i.e. activated proPO proteins) in the opsonizing serum would induce more rapid encapsulation reactions in the haemocoel than the saline-incubated particles, in a manner reminiscent of that described by Söderhäll et al. [69] for *A. astacus*. Moreover, in *C. destructor*, pre-immunization with low levels of LPS would raise host resistance to secondary infection with *Pseudomonas* CP [37] by triggering the proPO cascade in vivo, thereby liberating antimicrobial/agglutinating factors and priming the phagocytes within the host body (see below).

3.1.1 Microbial Killing and Cytotoxicity

Apart from its role in phagocytosis and as the mediator of immune adherence during capsule/nodule formation, the elements of the proPO cascade also participate in microbial killing and cytotoxicity. The fungitoxic action of quinones and melanin generated by PO activity from crayfish has been described by Söderhäll and Ajaxon [63], and more recently, we have obtained circumstantial evidence that, in crabs, the cascade mediates the killing of a Gram negative bacterium, *Moraxella* sp. in vitro [67]. The killing activity seems to reside exclusively in the proPO-bearing granular cells (Smith V.J. and Söderhäll K., unpublished), but is rarely detected in proPO preparations made from whole haemolymph [56]. In addition, Söderhäll et al. [70] have found that crayfish haemocytes display potent cytotoxicity towards ^{51}Cr-labelled human or mouse tumour and non-tumour cells in vitro, and while proPO preparations in crayfish HLS are ineffective in lysing the target cells, crayfish semigranular or granular cells depleted of their proPO systems by ionophore-induced exocytosis (Johansson M. and Söderhäll K., in prep.) lose their capacity to kill the tumour cell lines in culture [70]. In conclusion, activation of the proPO cascade will generate lytic factors directed towards the semigranular cells and these factors do not function towards mammalian cell lines. However, the exact role of the proPO proteins in cytotoxic reactions remains unknown.

The ability of invertebrate body fluids to destroy micro-organisms or other cells has long been considered an important part of the defence networks, and the lysis of erythrocytes has sometimes been taken to be indicative of complement-like activity in lower animals (see [9,14,19]). Consequently, a number of reports detailing the lytic properties of arthropod blood have been published (see, for example, [13,17,83], but outstanding amongst them is the work of Boman and his group in Sweden on the silkmoth, *Hyalophora cecropia* (see review by Boman and Steiner [10] and Chap. 6). In this insect, Boman and co-workers have shown that a family of powerful antibacterial proteins are synthesized by *H. cecropia* upon immunization of the pupae with bacteria [10]. These proteins, termed cecropins and attacins, have a low molecular weight but high hydrophobicity which permits them to selectively attack prokaryotic membranes [10]. The cecropins have now been purified and sequenced [26,72] (Chap. 6), and they appear to be produced within the fat body [18]. At present the relationship between cecropin production and the proPO cascade is obscure, but it is possible that components of the proPO system influence, in some way, the induction of the cecropins, or that the immune proteins modulate proPO activation in vivo. Certainly, melanin or its precursors are known to be highly toxic [20,87] (see also review by Menon and Haberman [39]), and administration of melanin inhibitors such as phenylthiourea favour survival of parasites within arthropod hosts [8,12,41]. Therefore, while it is likely that certain proPO proteins effect killing of invasive micro-organisms, it is essential that any potent toxic molecules generated by proPO activation are neutralized before they adversely affect the hosts' own tissues. With respect to crustaceans, there is no indication that lysozyme or lysozyme-like enzymes facilitate bacterial killing in vivo (McHenery, pers. comm.), so it remains to be established whether antibacterial activity is mediated by hydroxyl radicals or other reactive oxygen species in a manner akin to that shown by the pigment echinochrome A in echinoderms [44]. The cytotoxicity of *A. astacus* haemocytes towards tumour cells in vitro [70] resembles that previously described by Tyson and Jenkin [80] for *C. destructor* although, in the latter study, the trypsin treatment of the effector cells is likely to have caused release of the proPO proteins contained within them and resulted in loss of cytotoxicity. Until further research is undertaken it will remain a matter of speculation whether cytotoxicity and microbial killing are executed through the same mechanism.

3.1.1.1 Cell Communication

Perhaps, one of the most significant findings on the functional properties of the proPO system in crustaceans is its probable role in cell communication. A clue to this was provided by the finding, in crabs, that while phagocytosis is accomplished only by the hyaline cells [56] the proPO proteins which stimulate ingestion of foreign particles in vitro are enclosed by the semigranular and granular cells [65]. Other experiments have now ascertained that in crayfish, depletion of the granules from the semigranular cells abolishes their phagocytic potential (Söderhäll K., Smith V.J. and Johansson M., in prep.). The granular and, to a lesser extent, the semigranular cells of crustaceans therefore contribute the signals for the hyaline cells which stimulate greater phagocytic activity.

As mentioned above, only the semigranular cells of crayfish respond in vitro to LPS or $\beta 1,3$-glucans by exocytosis, thus releasing the proPO components (Johansson M. and Söderhäll K., in prep.). For the more stable granular cells (which contain greater amounts of the proPO enzymes) to undergo exocytosis in vitro, they have to be challenged with activated proPO proteins. Accordingly, this signal is proposed to be initially delivered primarily from the labile semigranular cells upon recognition of non-self within the haemocoel. Apart from being able to act opsonically and to induce exocytosis of granular cells, recent preliminary experiments have indicated further vital roles for the activated "sticky" proPO proteins in cell communication. They are able to trigger separated semigranular or granular cells to attach and spread or to aggregate (i.e. to form nodules) in vitro (Table 1). However, these active proteins remain

Table 1. Effect of the attaching proteins of the proPO cascade to mediate cell attachment and cell spreading of granular cells[a]

Coverslips coated with	Attachment	Spreading	Aggregation	Degranulation
Active proPO system	+	+	+	+
Inactive proPO system	–	–	–	–
5 mM $CaCl_2$ or 100 mM $CaCl_2$ and 1% BSA	–	–	–	–

[a] Coverslips were coated with an active lysate of crayfish haemocytes (homogenized in 5 mM Ca^{2+}, in which the proPO system is spontaneously activated) or with an inactive lysate (homogenized in 100 mM Ca^{2+}) for 1 h. The coverslips were extensively washed with H_2O, dried and a suspension of granular cells in 20 mM Ca^{2+} was added to the precoated coverslips. The behaviour of the cells on the coverslips was examined during 1 h. Data from Johansson and Söderhäll (unpublished)

Table 2. Biological functions connected with the proPO cascade or arthropods

Property	Freshwater crayfish	*C. maenas*	*G. mellonella*	*B. craniifer*
Opsonic	+ [57,69]	+ [57]	+ [48]	+ *a*
Clotting	+ [60]	+ [57]	NT	NT
Fungicidal	+ [63]	NT	NT	NT
Capsule/nodule formation	+ [69] and *b*	+ [59]	NT	NT
Lytic and degranulating factors	+ [58] and *c,d*	+ *d*	NT	NT
Chemotaxis	NT	NT	NT	NT

[a] Leonard, Ratcliffe and Rowley, J Insect Physiol (in press)
[b] Persson, Vey and Söderhäll, unpublished
[c] Johansson and Söderhäll, J Comp Physiol (in press)
[d] Söderhäll, Smith and Johansson, Cell Tiss Res (in press)

to be separated and characterized, and their individual or cooperative functions must be elucidated and the distribution of their putative membrane receptors on the different haemocyte types determined.

Accordingly, we have proposed a scheme, in a separate review, how cell cooperation/communication may operate in crustaceans and other arthropods in vivo [67]. However, the biochemical nature of the communicating signals and their receptors on the haemocyte surface await characterization, as does the extent of cell cooperation in other defence responses. A summary of the present knowledge of the biological functions of the proPO cascade is given in Table 2.

4 Summary and Future Perspectives

The prophenoloxidase activating system, as understood in crustaceans, is a complex enzyme system present in the granular, and to a lesser extent, the semigranular cells, which plays a major role in the recognition of foreignness by the host. Its biochemical features have been most extensively studied in *Astacus astacus*, where it is known to comprise a cascade of serine proteases together with a coagulogen and other factors that are specifically activated by non-self molecules (β1, 3-glucans or LPS). Functionally, the proPO cascade, provides opsonins, initiates capsule/nodule formation, participates in coagulation, facilitates microbial killing and mediates communication/cooperation between the different haemocyte populations (Table 2). Much remains to be learned about the biochemistry and biological role of the cascade, but from the information available so far, it appears to bear a certain resemblance to the complement pathway of higher animals and might constitute a primitive form. Other arthropod groups seem to contain a proPO system like that of crustaceans, but how the cascade relates either biochemically or functionally to the "lectins" described in arthropods is unclear. The sticky (attaching) proteins liberated through glucan or LPS activation of proPO could serve as agglutinating factors in the haemolymph. The existence of lectin molecules in arthropod fluids is well proven and these substances undoubtedly contribute to homeostasis in a number of ways (for example, by transporting or neutralizing hormones, nutrients, etc.) but the extent to which the host relies on lectins for non-self recognition and haemocyte activation is a mystery. Indeed, Yeaton [85] has stressed that their role in arthropod immunity is far from unequivocal. A thorough evaluation of proPO activation and lectin production is now of high priority for invertebrate immunologists.

Acknowledgement. The financial assistance of the Swedish Natural Science Research Council to Kenneth Söderhäll and the Royal Society to Valerie J. Smith is gratefully acknowledged.

References

1. Acton RT, Weinheimer PF (1974) Haemagglutinins; primitive receptor molecules operative in invertebrate defense mechanisms. In: Cooper EL (ed) Contemporary topics in immunology, vol 4. Plenum, London, pp 271-282
2. Anderson RS (1975) Phagocytosis by invertebrate blood cells in vitro: Biochemical events and other characteristics compared with vertebrate systems. In: Maramorosch K, Shope RE (eds) Invertebrate immunity. Academic Press, London, pp 153-180
3. Ashida M (1971) Purification and characterization of prophenoloxidase from the hemolymph of the silkworm *Bombyx mori*. Arch Biochem Biophys 144: 749-762
4. Ashida M (1981) A cane sugar factor suppressing activation of prophenoloxidase in haemolymph of the silkworm *Bombyx mori*. Insect Biochem 11: 57-65
5. Ashida M, Ohnishi E (1967) Activation of prophenoloxidase in the hemolymph of the silkworm Bombyx mori. Arch Biochem Biophys 122: 411-416
6. Ashida M, Söderhäll K (1984) The prophenoloxidase activating system in crayfish. Comp Biochem Physiol 77B: 21-26
7. Ashida M, Ishizaki Y, Iwahana M (1983) Activation of prophenoloxidase by bacterial cell walls or by B,1-3 glucans in plasma of the silkworm *Bombyx mori*. Biochem Biophys Res Commun 113: 562-568
8. Beresky MA, Hall DW (1977) The influence of phenylthiourea on encapsulation, melanization and survival in larvae of the mosquito *Aedes aegypti* parasitized by the nematode *Neoplectana carpocapsae*. J Invertebr Pathol 29: 74-80
9. Bertheussen K (1984) Complement and lysins in invertebrates. Dev Comp Immunol Suppl 3: 173-181
10. Boman HG, Steiner H (1981) Humoral immunity in *Cecropia* pupae. In: Henle W, Hofschneider PM, Kaprowski M, Moaloe O, Melchers F, Rott R, Sweiger MG, Vogt Pk (eds) Current topics in microbiology and immunology, vol 95, Springer, Berlin Heidelberg New York, pp 75-91
11. Bretting H, Kalthoff H, Fehr S (1978) Studies on the relationship between lectins from *Axinella polypoides* agglutinating bacteria and human erythrocytes. J Invertebr Pathol 32: 151-159
12. Brewer FD, Vinson SB (1977) Chemicals affecting the encapsulation of foreign material in an insect. J Invertebr Pathol 18: 287-289
13. Cantacuzéne J (1923) Le probleme de l'immunité chez les invertébrés. C R Hebd Seanc Soc biol (Paris) 75th Annivers 88 (Suppl): 48-119
14. Cooper EL (1980) Phylogeny of cytotoxicity. Endeavour (Oxf) 4: 160-165
15. Cornick JW, Stewart JE (1973) Interaction of the pathogen *Gaffkya homari* with the natural defence mechanisms of *Homarus americanus*. J Fish Res Boar Can 25: 695-709

16. Edelman GM (1983) Cell adhesion molecules. Science 219: 450-457
17. Evans EE, Painter B, Evans ML, Weinheimer PF, Acton RT (1968) An induced bactericidin in the spiny lobster *Panulirus argus*. Proc Soc Exp Biol Med 128: 394-398
18. Faye I, Wyatt GR (1980) The synthesis of antibacterial proteins in isolated fat body from *Cecropia* silkmoth pupae. Experientia (Basel) 36: 1325-1326
19. Finstad CL, Litman GW, Finstad J, Good RA (1972) The evolution of the immune response XIII. The characterization of purified erythrocyte agglutinins from two invertebrate species. J Immunol 108: 1704-1711
20. Graham DG, Tiffany SM, Vogel FS (1978) The toxicity of melanin precursors. J Invest Dermatol 70: 113
21. Häll L, Söderhäll K (1982) Purification and properties of a protease inhibitor from crayfish hemolymph. J Invertebr Pathol 39: 29-37
22. Häll L, Söderhäll K (1983) Isolation and properties of a protease inhibitor in crayfish (*Astacus astacus*) cuticle. Comp Biochem Physiol 76B: 699-702
23. Hergenhahn HG, Söderhäll K (1985) α_2 macroglobulin-like activity in plasma of the crayfish *Pacifastacus leniusculus*. Comp Biochem Physiol 81B: 833-835
24. Heyneman RA, Vercauteren RE (1968) Evidence for a lipid activator of prophenoloxidase in *Tenebrio molitor*. J Insect Physiol 14: 409-415
25. Hultmark D, Steiner H, Rasmuson T, Boman HG (1980) Insect immunity: Purification and properties of three inducible bactericidal proteins from the hemolymph of immunized pupae of *Hyalophora cecropia*. Eur J Biochem 106: 7-16
26. Hultmark D, Engström Å, Andersson K, Steiner H, Bennich H, Boman HG (1983) Insect immunity. Attacins, a family of antibacterial proteins from *Hyalophora cecropia*. EMBO J 2: 571-576
27. Inaba T, Suetake Y, Funatsu M (1963) Studies on tyrosinase in the housefly II. Activation of pro-tyrosinase by SDS. Agr Biol Chem 27: 332-339
28. Komano H, Mizuno D, Natori S (1980) Purification of a lectin induced in the hemolymph of *Sarcophaga peregrina* larvae on injury. J Biol Chem 255: 2919-2924
29. Lackie AM, Tackle G, Dularay B (1983) The non-recognition of negatively charged surface by locust haemocytes. In: Cooper EL, Wright RK (eds) Aspects of developmental and comparative immunology, vol 2. Pergamon, Oxford, p 54
30. Leonard CM, Söderhäll K, Ratcliffe NA (1985) Isolation and properties of the prophenoloxidase activating system in an insect *Blaberus craniifer*. Insect Biochem 15: 803-810
31. Liener IE (1976) Phytohaemagglutinins (phylolectins). Annu Rev Plant Physiol 27: 291-319
32. Lis H, Sharon N (1973) The biochemistry of plant lectins (phytohaemagglutinins). Annu Rev Biochem 42: 541-550
33. Marchalonis JJ (1978) Immunity in evolution. Harvard University Press, Cambridge, p 316
34. Marchalonis JJ, Edelman G (1968) Isolation and characterization of a hemagglutinin from *Limulus polyphemus*. J Mol Biol 32: 453-465
35. McCumber LJ, Hoffmann EM, Clem LW (1979) Recognition of viruses and xenogeneic proteins by the blue crab *Callinectes sapidus*: A humoral receptor of T_2-bacteriophage. J Invertebr Pathol 33: 1-9
36. McKay D, Jenkin CR (1970a) Immunity in the invertebrates: The role of serum factors in phagocytosis of erythrocytes by haemocytes of crayfish (*Parachaeraps bicarinatus*). Aust J Exp Biol Med Sci 48: 130-170
37. McKay, Jenkin CR (1970b) Immunity in the invertebrates: Correlation of the phagocytic activity of haemocytes with resistance to infection in the crayfish (*Parachaeraps bicarinatus*). Aust J Exp Biol Med Sci 48: 609-617
38. McKay D, Jenkin CR, Rowley D (1969) Immunity in the invertebrates I. Studies on the naturally occurring haemagglutinins in the fluid from invertebrates. Aust J Exp Biol Med Sci 47: 125-134
39. Menon IA, Haberman HF (1977) Mechanisms of action of melanins. Br J Dermatol 97: 109-112
40. Metchnikoff E (1884) Ueber eine Sprosskrankheit der *Daphnien*. Arch Pathol Anat 96: 177-195

41. Nappi AJ (1973) The role of melanization in the immune reaction of larvae of *Drosophila alonquin* against *Pseudocoila bochei*. Parasitology 66: 23-32
42. Noguchi H (1903) On the multiplicity of the serum haemagglutinins of cold blooded animals. Zentralbl Bakteriol I Abt Orig 34: 286-288
43. Paterson WD, Stewart JE (1974) In vitro phagocytosis by haemocytes of the American lobster (*Homarus americanus*). J Fish Res Board Can 31: 1051-1056
44. Perry G, Epel D (1981) Ca^{2+} stimulated production of H_2O_2 from naphthoquinone oxidation in *Arbacia* eggs. Exp Cell Res 134: 465-470
45. Preston JW, Taylor RL, (1970) Observations on the phenoloxidase system in the haemolymph of the cockroach *Leucophaea maderae*. J Insect Physiol 16: 1729-1744
46. Pye AE (1974) Microbial activation of prophenoloxidase from immune insect larvae. Nature 251: 610-613
47. Ratcliffe NA, Rowley AF (1979) A comparative synopsis of the structure and function of the blood cells of insects and other invertebrates. Dev Comp Immunol 3: 189-243
48. Ratcliffe NA, Leonard CM, Rowley AF (1984) Prophenoloxidase activation: Non-self recognition and cell co-operation in insect immuntiy. Science 226: 557-559
49. Renwrantz L, Stahmer A (1983) Opsonizing properties of an isolated hemolymph agglutinin and demonstration of lectin-like recognition molecules at the surface of hemocytes from *Mytilus edulis*. J Comp Physiol 149: 535-546
50. Rizki RM, Rizki TM (1979) Cell interactions in the differentiation of a melanotic tumour in *Drosophila*. Differentiation 12: 167-178
51. Rizki RM, Rizki TM (1984) Selective destruction of a host blood cell type by a parasitoid wasp. Proc Natl Acad Sci USA 81: 6154-6158
52. Rizki TM, Rizki AM (1983) Blood cell surface changes in *Drosophila* mutants with melanotic tumours. Science 220: 73-75
53. Rowley AF, Ratcliffe NA (1980) Insect erythrocyte agglutinins. In vitro opsonization experiments with *Clitumnus extradentatus* and *Periplaneta americana* haemocytes. Immunology 40: 483-492
54. Salt G (1970) The cellular defence reactions of insects. Cambridge University Press, London
55. Scott MT (1971) Recognition of foreignness in invertebrates II. In vitro studies of cockroach phagocytic hemocytes. Immunology 21: 817-827
56. Smith VJ, Ratcliffe NA (1978) Host defence reactions of the shore crab *Carcinus maenas* (L) in vitro. J Mar Biol Assoc UK 58: 367-379
57. Smith VJ, Söderhäll K (1983) B 1,3-glucan activation of crustacean haemocytes in vitro and in vivo. Biol Bull (Woods Hole) 164: 299-314
58. Smith VJ, Söderhäll K (1983) Induction of degranulation and lysis of haemocytes of the freshwater crayfish *Astacus astacus* by components of the prophenoloxidase activating system in vitro. Cell Tissue Res 233: 295-303
59. Smith VJ, Söderhäll K, Hamilton M (1984) B1,3-glucan induced cellular defence reactions in the shore crab *Carcinus maenas*. Comp Biochem Physiol 77A: 635-639
60. Söderhäll K (1981) Fungal cell wall B1,3-glucans induce clotting and phenoloxidase attachment to foreign surfaces of crayfish haemocyte lysate. Dev Comp Immunol 5: 565-573
61. Söderhäll K (1982) Prophenoloxidase activating system and melanization – a recognition system in arthropods – a review. Dev Comp Immunol 6: 601-611
62. Söderhäll K (1983) B1,3-glucans enhancement of protease activity in crayfish haemocyte lysate. Comp Biochem Physiol 74B: 221-224
63. Söderhäll K, Ajaxon R (1982) Effect of quinones and melanin on mycelial growth of *Aphanomyces* spp and extracellular protease of *Aphanomyces astaci* a parasitic on crayfish. J Invertebr Pathol 39: 105-109
64. Söderhäll K, Häll L, (1984) Lipopolysaccharide – induced activation of prophenoloxidase activating system in crayfish haemocyte lysate. Biochim Biophys Acta 797: 99-104
65. Söderhäll K, Smith VJ (1983) Separation of the haemocyte populations of *Carcinus maenas* and other marine decapods and prophenoloxidase distribution. Dev Comp Immunol 7: 229-239

66. Söderhäll K, Smith VJ (1984) The prophenoloxidase activating system – a complement-like pathway in arthropods. In: Aist J, Roberts DW (eds) Infections processes of fungi. Rockefeller Foundation Study Bellagio Italy, pp 160-167
67. Söderhäll K, Smith VJ (1986) The prophenoloxidase activating cascade as a recognition and defence system in arthropods. In: Gupta AP (ed) Humoral and cellular immunity in arthropods. Wiley, New York (in press)
68. Söderhäll K, Unestam T (1979) Activation of crayfish serum prophenoloxidase: The specificity of cell wall glucan activation and activation by purified fungal glycoprotein. Can J Microbiol 25: 404-416
69. Söderhäll K, Vey A, Ramstedt M (1984) Hemocyte lysate enhancement of fungal spore encapsulation by crayfish hemocytes. Dev Comp Immunol 8: 23-30
70. Söderhäll K, Wingren A. Johansson M, Bertheussen K (1985) The cytotoxic reactions of haemocytes from the freshwater crayfish *Astacus astacus*. Cell Immunol 94: 326-332
71. Starkey PM, Barrett AJ (1977) α_2 macroglobulin, a physiological regulator of protease activity. In: Barrett AJ (ed) Proteinases in mammalian cells and tissues. Elsevier/North Holland, Amsterdam, pp 665-696
72. Steiner H, Hultmark D, Engstom A, Bennich H, Boman HG (1981) Sequence and specificity of two antibacterial proteins involved in insect immunity. Nature 292: 246-248
73. Stephens JS (1962) Influence of active immunization on melanization of the blood of wax moth larvae. Can J Microbiol 8: 719-725
74. Stewart JE, Zwicker BM (1972) Natural and induced bactericidal activities in the haemolymph of the lobster *Homarus americanus*: Products of haemocyte: plasma interactions. Can J Microbiol 18: 1499-1511
75. Stoltz DB, Cook DI (1983) Inhibition of host phenoloxidase activity by parasitoid hymenoptera. Experientia (Basel) 39: 1022-1024
76. Taylor RL (1969) A suggested role for the polyphenol-phenoloxidase system in invertebrate immunity. J Invertebr Pathol 14: 427-428
77. Tyler A, Metz CB (1945) Natural heteroagglutinins in the serum of the spiny lobster *Panulirus interruptus* II. Chemical and antigenic relation to blood proteins. Biol Bull (Woods Hole) 89: 193-200
78. Tyson CJ, Jenkin CR (1973) The importance of opsonic factors in the removal of bacteria from the circulation of the crayfish (*Parachaeraps bicarinatus*). Aust J Exp Biol Med Sci 51: 609;615
79. Tyson CJ, Jenkin CR (1974a) Phagocytosis of bacteria in vitro by haemocytes from the crayfish (*Parachaeraps bicarinatus*). Aust J Exp Biol Med Sci 52: 341-348
80. Tyson CJ, Jenkin CR (1974b) The cytotoxic effect of haemocytes from the crayfish (*Parachaeraps bicarinatus*) on tumour cells of vertebrates. Aust J Exp Biol Med Sci 52: 915-923
81. Unestam T, Söderhäll K (1977) Soluble fragments from fungal cell walls elicit defence reactions in crayfish. Nature 167: 45-46
82. Vasta GR. Marchalonis JJ (1983) Humoral recognition factors in the Arthropoda: The specificity of Chelicerata serum lectins. Ann Zool 23: 157-171
83. Weinheimer PF, Evans EE, Stroud RM, Acton RT, Painter B (1969) Comparative immunology: A natural hemolytic system of the spiny lobster *Panulirus argus*. Proc Soc Exp Biol Med 130: 322-326
84. Yeaton RW (1981a) Invertebrate lectins I. Occurrence. Dev Comp Immunol 5: 391-402
85. Yeaton RW (1981b) Invertebrate lectins II. Diversity of specificity, biological synthesis and function in recognition. Dev Comp Immunol 5: 535-545
86. Yoshino T (1983) Lectins and antibodies as molecular probes of molluscan hemocyte surface membranes. Dev Comp Immunol 7: 641-644
87. Zlotkin E, Gurevitz M, Shulov A (1973) The toxic effects of phenoloxidase from the haemolymph of tenebrionid bettles. J Insect Physiol 19: 1057-1065

Subject Index